THE WORLD NATURALIST

The Natural History of the Dog

THE WORLD NATURALIST / Editor: Richard Carrington

The Natural History of the Dog

Richard and Alice Fiennes

I MURDERER: We are men, my liege.
MACBETH: Ay, in the catalogue ye go for men;
  As hounds, and greyhounds, mongrels, spaniels, curs,
  Shoughs, water-rugs, and demi-wolves and clept
  All by the name of dogs: the valu'd file
  Distinguishes the swift, the slow, the subtle,
  The house-keeper, the hunter, every one
  According to the gift which bounteous nature
  Hath in him clos'd; whereby he does receive
  Particular addition, from the bill
  That writes them all alike: and so of men.
                Macbeth, III, i

Weidenfeld and Nicolson
5 Winsley Street London W1

SBN 297 76455 1

Made and printed in Great Britain by
William Clowes and Sons, Limited, London and Beccles

To
Diana, Frances, and George

Contents

Plates

PLATES

Acknowledgements for photographs

Plates 8 and 20 are reproduced by gracious permission of Her Majesty the Queen. The publishers also acknowledge with thanks the permission of the following to reproduce photographs: the Trustees of the British Museum, plates 2–6, 9, 22; Mr Derek Taylor, plate 7; the Zoological Society of London, plates 10, 12, 13, 16, 24–35; Miss V. Tudor Williams, plate 11; Mrs Connie Higgins, plate 14; the Canadian Wildlife Service and Dr A.W.F.Banfield, plate 15; Major M.E.B.Banks, plates 17, 18; Dr Sylvia Sikes, plate 19; the Trustees of the National Gallery, plate 21; Mr H.Corbett, plate 23; Mr T.Dennett, plates 36–38, 40; the Ministry of Building and Public Works, plate 39. The photograph reproduced in plate 1 is by Richard Fiennes.

Figures

Foreword by Richard Fiennes

MY interest in the domestic dog as a subject of study was first aroused when I was persuaded by the editor of *New Scientist* to write an article on the dog for that journal. I had vigorously protested my unsuitability, but it appeared that there were very few – if any – people in the country with specialized knowledge on the subject, the late Professor Zeuner having died some months previously.

Up to that time, I had never really thought much about dogs as a problem of natural history. I had owned a number of dogs of different prescriptions, bred some, and trained them for use with the gun during the years I had worked in Africa. However, the dog as a scientific problem intrigued me so greatly after I had accumulated the material for the article which had been commissioned, that I continued to study it and to read all I could find on the subject. The present work, in which my wife has collaborated, is the result.

There are a great many books on dogs which make excellent reading and contain a wealth of information on the subject. We have attempted to do justice to these as sources of information by listing them in the bibliography. But few purely scientific treatments of the dog exist, other than studies on specialized aspects such as genetics. We have extracted the relevant information from these works in such a way as to present a picture of the domestic dog against a background of his natural history.

We should like to think of this book as being truly scientific, in the sense that we have collected our facts and checked them with care. We have then drawn from them those deductions which we consider to have validity; and in so far as possible, we have checked our deductions by original studies on skulls, teeth and other parts of the skeleton of both wild and domestic Canidae. In this way we have found some new facts not previously described.

However, we have avoided the form of a scientific monograph and hope that the book will be readily intelligible to the non-scientist who is interested in dogs. The conclusions we have reached differ in important aspects from those of other writers. We believe they provide the most probable explanation of how the dog came to be domesticated, by whom and why, from what ancestral forms, and of how so many breeds, of varying conformation and temperament, came into existence. As far as we know, no rational explanation of these matters has hitherto been advanced.

There may be many who disagree with our deductions. Others may put different interpretations on the same facts, or new facts may come to light on which to base revised judgments. If so, we shall not be dismayed; on the contrary, we shall think that our work has been worth while.

We have received from the publishers the usual courtesy, which I have come to expect in my dealings with them, and I express my appreciation particularly of the help given by Richard Carrington, Nicolas Thompson, and Julian Shuckburgh. We are also grateful to Mrs Sonia Cole for her help over the final revision of our typescript.

The pictures of Queen Victoria's Pekingese 'Lootie', and that of George Stubbs' 'Fino and Tiny' are reproduced by gracious permission of Her Majesty the Queen.

Plates 2–6, 9, and 22 are reproduced by permission of the Trustees of the British Museum, and we express our thanks particularly to the Keeper of Oriental Antiquities and to Mr A.F.Shore of the Department of Egyptian Antiquities for their help in finding us suitable material.

Plate 21 of Gainsborough's 'Pomeranians' is reproduced by courtesy of the Trustees of the National Gallery.

Plate 39 of the Windmill Hill dog is reproduced by permission of the Ministry of Building and Public Works. We are especially grateful to Major H.L.Vatcher and Mrs F.deM.Vatcher, Curators of the Alexander Keiller Museum at Avebury, who not only provided the print, but permitted us to make a detailed examination of all canine material in this Museum.

Canine skulls have been loaned to us by the Department of Mammals of the British Museum (Natural History) and we express our particular thanks to the Trustees and to Dr Frazer for his help.

A number of plates (10, 12, 13, 16, and 24–35) were obtained from the photographic library of the Zoological Society, and we give acknowledgement for permission to reproduce these to the Society. We are also grateful to Mr Fish, the Society's Librarian, for his help in selecting suitable material.

Plate 15 of the wolf rounding up Caribou was supplied by the Canadian

Wildlife Service. The photograph was taken by Mr A.W.F.Banfield of this service and supplied by Mr Darrell Eagles, Head of the Editorial and Information Section. To both these gentlemen we express our thanks.

Plate 11 of the basenji dog was supplied to us by Miss V.Tudor Williams and to her also we express our gratitude.

Plate 14 of the Middle East feral dog 'Shebaba' was supplied by Mrs Connie Higgins and we are grateful to her both for giving us the photograph of her bitch and for co-operation in our studies of this breed.

Plates 17 and 18 of the Greenland huskies were supplied by Major M.E.B.Banks, and we are grateful to him for permission to reproduce them. We are also grateful to Dr Sylvia Sikes for the photograph of her elkhound bitch, reproduced in plate 19.

Plate 23 of a mongrel dog is reproduced by permission of Mr H.Corbett, Senior Technician of the Zoological Society. We are most grateful to him for supplying the photograph and for permission to reproduce it.

The silhouettes and photographs of skulls of Canidae were taken by Mr Terry Dennett, Laboratory Technician to the Zoological Society, and we are appreciative of his excellent work. Plate 7 of the authors' 'Fo Dogs' is reproduced from a photograph made by Mr Derek Taylor of the Nuffield Institute of Comparative Medicine, to whom we also express our gratitude.

May 1968

Zoological Society,
Regent's Park,
London

Part 1

Dogs in History

Chapter 1

Dog and Man

IT is usually said that the first animals to be domesticated were either reindeer or dogs. Dogs were undoubtedly first, because the more primitive races of man who keep reindeer are unable to do so without the aid of herd dogs. The process of domestication is unlikely to have been a sudden or deliberate one, as could have been the case, say, with cattle or horses.

Twenty thousand years ago there existed some association between two groups of animals, man and wolf; both were predators on vast herds of grazing mammals which roamed the steppes and arctic tundra. Supplies of food were more than adequate for both and each hunted in his own way. In many respects the wild canine stocks had attributes which man lacked; but man possessed intelligence, cunning, and a power to organize which was superior to that of the wolf. It is known from archaeological evidence that Palaeolithic man at times massacred wild horses and other animals in numbers far beyond his needs by driving them into traps, or over cliffs. One may suppose that wolf packs would come to follow man's trail and would share in the feast of discarded meat. In this way certain groups of wolves would become partially dependent on man for their sustenance and would follow him on his forays.

To suppose this is more than just speculation, since we well know that in times of scarcity all members of the genus *Canis* take readily to scavenging; and indeed some groups obtain their requirements more in this way than in any other. It is thus clear that, initially, dog – or rather wolf – adopted man rather than that man, by a deliberate act, captured and reared young wolf cubs as domestic animals.

While certain canine stocks may in early times have become dependent on man, it is unlikely that man had any special love of his rival wolves. No doubt on occasion wolves would attack stragglers or seize young children, and man was probably to some extent in awe of his rival predators. This would account for the old superstitions about werewolves, or fables such as that of Romulus and Remus suckled by the she-wolf, and for similar stories from India. Wolves were first domesticated because of their usefulness at a time of climatic change when the arctic tundra, supporting vast herds of grazing animals in open country, gave way to thick forests which man feared to enter and to which his hunting skills were not adapted. Animals prepared to become man's ally, skilled in hunting by scent in thick country, would be a welcome asset. Furthermore the dog's ability to round up animals and drive them into traps and corrals could be the essential forerunner of domestication.

Among more advanced people domestication, in the sense of selective breeding of dogs for specific purposes, made significant advances between Palaeolithic and Mesolithic times, that is between 20,000 and 10,000 years ago. Man, adapting himself to the new climatic conditions which appeared in Mesolithic times, found that his hanger-on, the wolf, possessed valuable properties complementary to his own, of which he could make use and which assisted him in entering into a new way of life which new conditions forced on him.

Recognizable breeds of dogs have been recovered from archaeological sites in connection with Mesolithic settlements, particularly in the Maglemosean deposits in Denmark and in the lake settlements found in Switzerland, Austria, and elsewhere. Sheep dogs, hunting dogs, and miniature dogs of the Maltese type were already present during the period when the Mesolithic was gradually being replaced by the Neolithic.

By the time that written history appears, all the main groups of dogs were in existence and can be recognized from representations in stone and pottery from Egyptian, Assyrian, and Greek sources. Dog breeding is an ancient art and modern breeders have contributed little to what was already done by the end of the Neolithic period.

Today, we wonder at the almost incredible diversity we find in domestic dogs, the immense differences between the various breeds. We are amused by the ludicrous sight of a great dane greeting a diminutive Yorkshire terrier in canine fashion with bounding, joyous abandon. We know these animals are all dogs – even a child knows this – and so do the dogs themselves. Yet scientifically we see that differences in the group are such that, in other genera, taxonomists might have little difficulty in indulging in their favourite sport of creating new species.

Both Linnaeus and Buffon attempted specific classifications of the domestic dog based on conformation and colour, but later workers have not recognized such speciation as valid, and rightly so. Modern dogs are so interbred that to divide them into species would, in any case, not be

feasible. Furthermore, many of the features which might be used in classification have been introduced artificially by man by selective breeding,
not by natural selection and adaptation to the dictates of habitat and way
of life. Thus many modern breeds would be unable to survive under wild
conditions, although the more primitive breeds can do so and in many
instances have become secondarily wild, or to use the technical term 'feral'.

When studied in detail the differences of conformation among dogs,
though striking to outward appearances, are not so fundamental as might
be supposed. They consist mainly of size, length of leg; length or shorten-
ing of the muzzle; size and attitude of the ears; length, shape, hairy covering,
and carriage of the tail; and differences of the pelage, that is the length and
character of the hair – whether curly or straight, wiry or silky, and so on.

A number of characteristics retained in the adult are essentially those of
puppies and may be ascribed to the phenomenon known in genetics as
'neoteny', that is, a reversion to primitive characteristics found in the young
of the species. In all wild canid stocks, ears are straight and erect, whereas
in puppies they are floppy. The coat of puppies is silky, not wiry; and the
tail at rest hangs between the legs, covering the anus. In man himself,
features characteristic of neoteny are very marked; one example is the
absence of marked supraorbital ridges, which were present in his forebears.
The human skin has also been likened to foetal skin, supposedly due to an
arrest of development.

In size dogs vary from the four to five pounds of the chihuahua to 180
or so pounds in the more massive breeds. This again is striking; but both
absolute size and leg length are very variable in all groups of animals. In
man, cattle, buffalo, elephant, and even in crocodiles, large and pigmy
races are known.

Variations of size soon develop even under natural conditions as a result
of adaptation to environment. Creatures which live in desert or plains and
which must travel long distances in search of food and water tend to
become long, lean and long-legged, whereas some forest living creatures
become smaller than their prototypes.

Almost every group of dogs has produced its toy or pet breed, and 'lap'
and 'sleeve' dogs have been developed in almost every country of the
world. The 'toy' spitz dog, the so-called Maltese, has been a favourite
pet with women since early Graecian times and throughout the Roman era.
The pekingese and pug types of miniature dogs were developed in the
orient at least 2,000 years ago. All have survived until today and many
more miniature breeds have been developed from every group.

The main basis of classification within a genus of animals must ultimately

5

rest not on such superficial differences, however striking, but on anatomical differences, which can be recognized not only in recent types, but also in fossil specimens. In this respect, all the Canidae are remarkably uniform and anatomical differences are slight even between genera, let alone species. Differences of temperament and way of life are rather more marked, but even here this family is uniform; when pressed by physical circumstances wolves, for instance, will feed and act like foxes, and jackals in some places will combine in packs to hunt like wolves.

The Canidae were derived from more primitive ancestral predators in rather recent times (late Pliocene and early Pleistocene), and radiated rapidly into different habitats without losing the ability to adapt themselves to different conditions. Therefore their genetical make-up has retained for them the power, when pressed, to acquire a living in many varied ways, even by subsisting on vegetarian diets, either entirely or partially. Thus the family has in it *ab initio* a power of variability not found in most groups of animals. The genus *Canis*, perhaps alone among mammals except for some primates and especially man, possesses variable genetical features by which members even of a single litter are born with marked differences of conformation and temperament.

The Canidae have pronounced instincts for territory and hierarchy and have a complicated social etiquette. Instincts of territory and hierarchy exist almost universally in Mammalia and indeed in much more primitive animals, even in fish. Among wolf stocks, these instincts have advanced to the stage where there is social organization and division of labour. We believe that among the wolves there exist genetical variations, even in members of the same litter, which separate them into distinct groups, having different abilities as regards their hunting habits.

Such pre-existing genetical tendencies would make it easy for primitive man to breed from wolves dogs with herding instincts and establish in a short time a breed of shepherd dogs. Similarly, he would have little difficulty in establishing other breeds with a highly developed sense of scent and strong hunting instincts. These would eventually produce the great groups of hunting dogs, used by man since time immemorial. The high genetical variability of *Canis*, supplemented by breeding fox or jackal blood into the group, could readily account for the strange diversity of form and temperament so characteristic of our dogs today.

In 20,000 years man has developed from a nomadic predator animal to a superb master of technology, contemplating landing on the moon and exploring the solar system. So rapid has been this advance that there still exist in isolated pockets races who live – or were living until recently –

lives similar to those of our not so remote ancestors who set in motion this incredible revolution. The early relationship between man and dog may be compared with that which exists between the aboriginal inhabitants of Australia and their dingos, described in a later chapter. It can still be seen in the relationship between the Eskimo peoples of the north and their primitive huskies, which they use for hunting and to draw their sledges. No attempt is made by the aborigines to improve the dingo stock by selective breeding; and most of them live, breed, and hunt in the wild, only a few being brought into captivity and reared in domestication. Some attempts are made sporadically by the Eskimos to breed from those dogs which show aptitude for sledge work, but this is haphazard and every few generations the husky stock is bred back to the parent wolf by tethering bitches on heat where the dog wolves will find them. Thus, the evidence for a gradually developing association receives strong support from observations of still existing primitive people.

The Australian aborigines are said to treat their dingos with a degree of affection; the same cannot be said of the treatment of their huskies by Eskimo peoples. This point is forcibly made in an illuminating article by C.G. and E.G.Bird [4], from which the following extract is taken.

'We have seen Eskimos take a disobedient dog and tie a knot around the dog's neck and pull the ends until the animal was senseless. Also in order to stop the dogs wandering, they tied up one leg. This might be all right in a temperate country, but in the Arctic the limb freezes.'

These authors also quote from Croft, who hung his dogs till they were senseless in order to knock the tips off the back molars of the bottom jaw in order to prevent them from eating their harness. Edward Shackleton wrote: ' . . . and the number of whip handles which we broke on the cast-iron heads of our dogs was very great.' Reluctantly we must assume that the barbaric practices here described towards polar dogs in the belief that they were 'merely' wild animals were typical of those used when man first trained wild wolf stocks.

Dogs began a life of domestication as slaves rather than allies, and the warm relationship which has since developed – and indeed was in existence in early dynastic times in Egypt – developed gradually, together with mutual understanding and regard. It is a truism to say that the dog is largely what his master makes him: he can be savage and dangerous, untrustworthy, cringing and fearful; or he can be faithful and loyal, courageous, and the best of companions and allies.

The literature from ancient times up till the present is disappointingly unhelpful in suggesting the origins of dogs. However, some survey of the early writers will be helpful in developing certain points which must be established. This literature has been ably summarized by Ash [2] and the following account is taken largely from his pages.

Xenophon (*c.* 430 – 355 BC), an early pupil of Socrates, after a colourful life as an Athenian historian, writer, general, renegade, and traitor, retired in old age to Scillus in Elis where he lived as a country gentleman and wrote books. Among his other occupations, he became an enthusiastic breeder of dogs, of which he wrote in his *Cynegetica*. In this he defines two kinds of dogs, the castor-dog and the fox-dog. Castor-dogs were used in packs for hunting hares, and from his description could well have been in the nature of beagles. Evidently they hunted by scent and would go off in full cry, as do our hounds today, with a great yapping and baying. The 'fox-dog' was so-called because this animal sprang from a union between dog and fox and in the course of time the two strains became confused. Xenophon had a very poor opinion of these dogs and regarded them as virtually worthless. His statement that they were derived from crossing dog and fox, although it cannot be regarded as scientific evidence, cannot be dismissed outright; it raises the possibility that in establishing canine breeds, stocks other than those derived from the basic wolf blood have been used.

Xenophon describes in detail his view of the perfect hunting dog, in which he himself was most interested. Although he distinguishes only two breeds of dogs, he was obviously acquainted with others. He recommends the use of Indian dogs for hunting fawns and stags: 'for they are large, strong, swift of foot and spirited, and having those qualities they are able to endure hard tasks.' For wild boar hunting he recommends Indian, Cretan, Locrian, or Spartan dogs. He states that 'one must avoid selecting from these breeds any or every sort of dog, for they must be ready to fight this savage animal. The hunters first advancing where they suppose the boar to be, should lead up the pack of hounds and let loose one Laconian (Spartan) dog, but the rest should be kept tied up.'

Aristotle (384 – 322 BC) wrote about dogs in his *Historia Animalium*. He also believed that dogs were mated with wild animals and remarks that in Cyrene wolves mate with dogs and that the Laconian dog is a hybrid of a fox and a dog. Bitches were said to be taken into the desert and tied there; many were devoured unless the wild beast was in the mood to mate. Elsewhere he says that the dogs of India result from a cross between a tiger and a dog. This belief is of course pure superstition; but the known practice in many parts of the world of tethering bitches where they could be covered

8

by wild dog wolves suggests that part of this story may not be without foundation. Indian dogs, notorious for their savagery, may have been crossed at intervals with the Asiatic wolf (*Canis lupus pallipes*) or with the dhole (*Cyon javanicus*); no doubt some of the unfortunate bitches so tethered would be devoured by tigers, which may have given rise to the superstition. These stories again suggest that outcrossing with exotic canine species may have occurred in early times, so contributing to the great capacity for variation in domestic dogs.

The elder Pliny (AD 23 – 79), in his great *History of the World*, mentions the habit of the Colophonians and Castabaleans of maintaining squadrons of mastiffs for war service. They were put into the front rank in battle and were never known to draw back and refuse to fight. He also describes episodes when dogs fought lions and elephants, and a specific occasion when dogs were crucified by Romans because they failed to bark when the Gauls scaled the Capitol. He tells us that the Romans thought the flesh of suckling whelps a particularly fine meat and used these animals for sacrifice and expiation.

The learned Roman writer Varro (82 – 27 BC), a great landowner, wrote of dogs in his well-known work *De Rustica*, in which he gives some useful practical notes on the keeping of dogs and on the breeds which existed in his time. Dogs, he says, should be kept to protect sheep and goats from wolves. There were two kinds of dogs, one for hunting the wild beasts of the woods, and the other trained for purposes of defence and used by shepherds. They should be of a handsome shape and great size, their eyes black or yellowish, with nostrils to match; the lips should be blackish or red. The head should be large, and the ears broad and hanging. These descriptions clearly indicate a shepherd dog of the type of the Pyrenean mountain dog. He prefers a dog that is white, as is the Pyrenean, because it is more easily seen; and the large hanging ears serve to identify dogs of this type. He states that dogs are called after the district from which they come, as Laconian, Epirote, Sallentine; and he also gives a great deal of practical information on feeding and breeding.

The Roman writer Arrian, who lived in the 2nd century AD, wrote a book on hunting which he called *Cynegeticus* in imitation of Xenophon, whom he both aped and criticized. He remarks condescendingly that what Xenophon left out, Arrian would supply and adds that Xenophon did not make these omissions deliberately, but because he knew nothing of Celtic dogs. These, states Arrian, 'have no equal in speed so that no hare can escape them.' Some of the Celtic dogs were called 'Segusii', a name he supposes to be taken from some Gallic tribe; these dogs, he says, 'are shaggy

9

and ugly to look at; they do not bark, and they whine in a miserable manner. The quick Celtic dogs are called in Celtic "Vertragi", a name given to them on account of their great speed.'

From this description, we can recognize the ancestors of the Irish wolfhound, known to be the main group of dogs kept by the Celtic peoples.

Oppian, a Greek who lived in Syria and wrote in the reign of the Roman emperor Caracalla towards the beginning of the third century AD, produces further interesting information about dogs.

'There is also a certain breed of dog, in this case used for tracking purposes – small indeed but well worth the tribute of our song (he wrote poems) – which the wild tribes of painted Britons are accustomed to breed, and to which they have given the special name 'Agassaei'. In size this dog is about equal of those worthless, greedy house (or pet) dogs that attend at table. It is round in shape, very skinny, with shaggy hair, and a dull eye but provided on its feet with deadly claws, and it has rows of sharp, close-set teeth, which contain poison. In powers of scent the 'Agassaeus' is easily the superior of all other dogs, and the very best in the world for tracking. Since it is very clever at finding the track of those creatures that walk the earth, it is also able to indicate with accuracy even the scent which is carried through the air.'

Unfortunately Oppian fails to give a full description of these dogs; but they must surely be the ancestors of our present breeds of terriers, even if there is a suggestion of 'pointing'. In Roman times Britain was famed for its export of dogs; they were usually described as mastiffs, but were very probably terriers.

Aelian, who wrote in Greek but lived in Italy (*c*. late second and third century AD), describes Indian dogs and repeats Aristotle's story that they were occasionally mated with tigers, which accounted for their extraordinary ferocity and fearlessness. Aelian was much concerned with the dog's devotion to his master and gives some interesting examples of this.

'When Darius, the last of the Persian kings, was killed by Bessus in his battle with Alexander, and lay dead, all the men left the corpse behind but the dog alone he had bred remained faithful. The dog belonging to King Lysimachus chose to die by the same fate as his master, although he could, had he so wished, have saved himself. Again, when there was civil war in Rome, a Roman citizen called Calvus was killed. Many of his enemies strove in rivalry to accomplish the glorious deed of cutting off his head, but none could do so until they had killed the dog who stood by his side.'

In relation to hunting, Aelian remarks that

'a good hunting dog as soon as he has caught his prey is glad, and treats it as a prize; he will not partake of the prey until his master arrives and makes such disposition to him as seems right. If he happens to find a dead hare or boar he will refuse to touch it, proving an innate sense of honour in the dog which prevents him eating anything that he does not catch. The hunting dog leads on the huntsman tied by a long strap, nosing about and not uttering a sound as long as there is no quarry and he comes across nothing in his path. Once he comes on a scent he stops until the huntsman comes up and the dog, overjoyed with the good luck, fawns on his master and caresses his feet; then he recommences on the track and keeps at walking pace until he arrives at the lair, not proceeding further. The huntsman whistles up his men and they cast their nets about the lair, and at the right time the dog barks with the intention of arousing and exciting the animal so that it may fall into the trap, and will then be caught by the nets. After the capture, the dog utters a glad cry of victory and exults and jumps about "like armed men when they have routed their enemies". In the case of stags and boars that is how the hounds behave.'

This description is of great interest and shows that methods of hunting developed in Norman and later times had their origin in very early days, and even that the prototypes of the retrievers, which capture a wounded bird without harming it, were well known in Roman times.

The Roman writer Grattius, who lived at the time of Augustus and wrote a poem *Cynegeticus*, gives some rather more informed details about the dogs of his day.

'Dogs have innumerable countries from which they spring and the disposition of each kind corresponds with his origin. The Mede dog shows great fight, though untaught, and the far distant Celt is celebrated with high renown. On the contrary, the Geloni will not fight and hate war; but they are naturally of a good scent; the Persian is endowed with both qualities.

'Some prefer to breed Indian dogs, a race of implacable anger, with which contrast those of Arcadia, that are tractable yet combative. The Hyrcanian (those near the Caspian) race of dogs have all the ferocity and more, for they interbreed with the *savage monsters of the forests*' [our italics].

'But the Umbrian dog runs away even from the enemies whom he has himself discovered. Would that he had as much courage and pluck in fight as he has loyalty and sagacity of scent! What if you were to go to the English Channel, surging with treacherous sea, and reach as far as the Britons themselves! How small the charge and expense if you do and are not attracted merely by the deceptive look and form (this is the only danger about British dogs): nay, when a great work is to be done and courage displayed and the hazard of approaching war gives the final summons, then you would not admire even the well-known Molossian hounds so much as these.

'With the British dog one may compare the clever Thessalian, that comes from Azarus or Pherae, and the wily Acarnanian; for as the Acarnanes steal secretly into the fight, so that hound cometh upon its enemies without warning. But the hound sprung from Aetolian stocks starts the boars he has not yet seen by his shrill bark: whether it is fear that causes his cry to be uttered or an over-quick haste and passion. And yet considering all a dog's accomplishment you would not rightly despise that breed: for they are marvels of speed and of quickness in scent; and there is no labour so heavy that it conquers them or makes them yield. So it would be my policy to intermix the breeds of dogs. An Umbrian dam will give to the slower-witted Gaul a lively intelligence; the Gelonians will inherit courage from an Hyrcanian sire; and the Calydonian having its vices corrected by a Molossian father will get rid of its great defect, a foolish tongue. So do we cull something from every flower while kindly nature seconds our efforts.'

Grattius' account of British dogs supports that of Oppian. He again refers to the unattractive form and appearance of British dogs, but remarks on their great courage and resource. Although he compares them with the much larger Molossian mastiffs, we may again assume that these were rather unremarkable dogs in appearance, probably ancestral terriers.

We have seen that all domestic dogs, classed as *Canis familiaris*, are not necessarily derived from a single stock – indeed the evidence points otherwise, and the testimony of early writers reviewed above may suggest that the parent stock, undoubtedly wolf, has also been crossed with jackal and other wild canine blood.

The difficulties of classifying domestic dogs from very early times will already be apparent. But in spite of these difficulties and despite the crossings between different groups which have occurred, we propose that modern dogs can be classified into four main groups, indicating four separate origins. We shall survey them under these headings:

1. THE DINGO GROUP derived from the pale-footed Asian wolf (*Canis lupus pallipes*) – a very homogeneous group of dogs which breed largely true to type;
2. THE NORTHERN GROUP derived mainly from the northern wolf (*Canis lupus*), comprising the huskies, samoyeds, chows, pomeranians, elkhounds, collies, alsatians, corgis, schipperkes and terriers;
3. THE GREYHOUND GROUP derived from a cursorial wolf ancestor, related to *C. l. pallipes*;
4. THE MASTIFF GROUP derived from mountain wolves such as the Tibetan wolf (*C. l. chanco* or *laniger*) and capable of transmitting certain rather special characteristics which are retained in

many derived breeds such as the spaniels, setters, pointers, retrievers and true hounds, as well as the mastiffs and bulldogs.

In addition to this classification, there are certain breeds of dogs, particularly in the American continent, which have been derived from local wolf stocks such as the coyotes, and which do not fall naturally into these groups.

We shall survey these four groups systematically. But meanwhile it is necessary to discuss the origins and dispersal of our domestic dogs, and the dogs of ancient and classical times.

Chapter 2

The Origins and Dispersal of Dogs

BEFORE proceeding to an account of the relationships between Stone Age man and wolves which led eventually to domestication, it is necessary to make some study of the wild stocks from which the domestic dog originated (a systematic survey of wild Canidae is given in Appendix 1).

Wolves have been classified under innumerable varieties, mostly based on rather trivial features of size and coat colour. These races were excellently reviewed by the late R.I.Pocock, FRS in 1935 [49]. Here we are concerned with four races only: the type species *Canis lupus*, the northern grey wolf; *Canis lupus pallipes*, the pale-footed Asian wolf; *Canis lupus arabs*, a race newly designated by Pocock for the small desert wolf of Arabia; and *Canis lupus chanco* or *laniger*, the woolly coated wolf of Tibet and northern India.

In spite of much interbreeding between the races of domestic dogs, these four wolf groups correspond to the four groups of domestic dogs which we have proposed, namely the northern (including the spitz) group; the dingo group, including the pariahs; the greyhound group; and the mastiff group.

In 1941, Pocock [50] made a further invaluable contribution to the study of wild Canidae in which he mentions these four wolf groups. In this work he writes:

'A point of interest connected with the wolf is the certainty of the species being the principal, if not the sole, ancestor of domestic dogs. There is no unanimity of opinion on this subject, some authors thinking that the jackal contributed to the strain thus accounting for the diminutive size of some of our breeds. That view cannot be summarily dismissed considering that some of the earliest known dogs possessed by Neolithic man were comparatively small. It is, moreover, generally believed in India that the jackal interbreeds with the pariah, and that in South America imported dogs sometimes cross with the more remotely related wild dogs of the country. But in view of the remarkable plasticity of the organization of Canidae, as attested by the modification of nearly all the external features in our breeds from historic times, it seems unnecessary to introduce the jackal as a factor in the case. Millar, indeed, excludes that species from the stock because of the presence on the upper molars of the cingulum, which is at all events less well-developed in domestic dogs. Although this is a somewhat elusive feature I am inclined to agree with him that it is needless to look beyond the wild prototype. There is at all events no mistaking the stamp of the wolf on such breeds as the alsatian and the Eskimo. Practically the only constant difference in enabling a wolf skull to be distinguished from a dog skull of the same size and shape lies in its heavier dentition. Domestication through long ages has no doubt reduced the size of the teeth in our dogs.'

Much has been made of the supposed differences between the teeth of jackals and those of wolves and dogs, particularly by Wood-Jones [30] and repeated by G.M.Vevers [63]. However, as Pocock states, these features are indeed elusive and we have not been able to establish from our studies of jackal, dog and wolf skulls that they do in fact exist. In any case, the presence of a cingulum (ridge) could not be used to exclude the jackal from the ancestry of the domestic dog unless it were known that this feature was a genetically dominant characteristic which would be inherited in hybrids.

Pocock goes on to give evidence that the northern wolf shades indistinguishably into the Tibetan wolf, the Tibetan wolf into the pale-footed Asian wolf, and the Asian wolf into the desert Mesopotamian wolf. He points out that the latter is little differentiated in size or other features from the larger jackals of the species *Canis aureus*. All of these animals can interbreed with each other and produce fertile offspring.

There is therefore no reasonable doubt that the ancestors of domestic dogs were wolves; but at some stage there may well have been some admixture of jackal blood. The wolf ancestry needs to be established authoritatively from the outset and further evidence will appear in our descriptions of the four canine groups. In this chapter we are concerned mostly with studying the habits and characteristics of wolves in the wild, so that their qualities good and bad may be understood. In this way we shall gain an insight into the behavioural characteristics which they have bequeathed to domestic dogs and into the problems which went into selecting them for different uses.

Much useful information on this subject is contained in the two works by Pocock already mentioned. However, by far the most extensive studies of the habits of wolves are contained in *The Wolves of North America* by Stanley P.Young and Edward A.Goldman [69]. So extensive are the references and anecdotes contained in this book that justice cannot be done to it here, though a few extracts will be quoted to make certain important points.

The northern wolf, which is the chief subject of study by these authors, is the timber wolf, an offshoot of the northern grey wolf of Europe and Asia. It exists on the North American continent in a variety of local races differing in size and colouration, which varies from white through grey and brownish-grey to black. Its preferred prey are the buffalo (that is, bison), antelope, elk, deer, caribou and moose in that order. These wolves also feed on domestic livestock, such as cattle, sheep and horses, and they constitute a serious pest on this account and because they reduce the numbers of wild animals on which the Indians and Eskimos subsist.

An old buffalo hunter, Charles Aubrey, quoted in the book, gave the following description of wolf tactics in the presence of a large buffalo herd.

'Bulls were on the flank and made up the rear guard of the column. The old cows were in the van and within the flank next to the bulls, while the young stock were distributed generally through the centre. With the herd also marched the beasts of prey which fed on the buffalo. Of these the chief gathering were the great Buffalo wolves, each the very incarnation of destruction. With his powerful jaws of sharp teeth, his wonderful muscular strength, the tireless endurance of a compact body, the speed of a greyhound, and the cunning of man ... the Grey Wolf must eat and that the best in the land, for he is no scavenger of the plains.

'The Indian was not the wolf's superior as an expert hunter and in concerted action in the attack upon prey, the buffalo, much as the cowboy cuts out from the herd the animal he has chosen, so the wolves selected their victim. With deceptive sleepy gait they closed in on the flanks of the marching host; when the leader had picked out his victim – preferably a young cow – he at once changed his gait to a quick pace, and his followers, alert to imitate his movements at once closed in. When he saw that they were well in hand he gave the signal for attack, a deep, hoarse roar and bounding rush followed. The terrified cow was cut out of the herd, once out the powerful leader made a quick, sidelong swing and hamstrung the prey, and the others as powerful fastened on her flanks.'

Other instances are given of the wolves dividing into two groups, one to separate the prey and the other to attack. When running down prey it is told how some of the wolves chase the animal in a circle, bringing it back towards the others which rest; then the first group rests while the second group takes over in the chase. It is also said that, as in the hunting tactics of Palaeolithic man, wolves would stampede herds of animals over cliffs, or drive them into rocky gorges where they would be trapped.

Wolves vary their tactics according to the characteristics of the animals they hunt; for instance with antelopes, which are inveterately curious animals, they adopt manoeuvres to excite their curiosity and induce them to investigate in the direction where the wolf pack lies in wait. In the case of large animals, one or two wolves would attract the animal's attention by attacking from the front, where they would seize on the lolling tongue or the nose and gouge out the eyes by savage bites. Once the animal's attention was distracted in this way, the main attack would come from behind and attempts would be made to sever the achilles tendons, thus rendering the animal powerless to move. Once helpless, all the wolves would fall on it and tear it to pieces.

When game is plentiful, these wolves will kill for the sheer lust of killing, only consuming a small portion of the carcass and then leaving it. However, when game is scarce, they will pick all flesh from the bones and sometimes bury a portion of the carcass or the bones beneath the ground, returning later to retrieve them, just in the same way as domestic dogs. When other food is plentiful they will not feed on carrion, but in times of scarcity they will follow human hunters, feeding on the offals which they leave. They will also unearth and eat buried human cadavers. These two characteristics give an insight into the early associations between man and wolf. Wolves are also fond of some kinds of vegetable foods and at certain times of the year, feed almost exclusively on wild fruits, including a certain kind of plum. Their feeding habits are said to be most irregular. They prefer to feed on large prey and after the kill they will gorge to such an extent that they become drunk with meat and are then easily approached and shot.

Wolves show great concern and affection for each other, as is shown by the following anecdote related by a hunter named Freuchen, who observed a family of wolves north of Hudson Bay. He states:

'As soon as we came ashore we heard the howling of wolves quite close, about 100 metres away. We went to look for them and saw five. Two old white wolves and three cubs that were nearly as large as their parents, but grey. They were sitting on the ground, noses in the air and howling, but sprang up when they saw us and fled. Shortly afterwards we caught sight of a fourth cub, grey like the others, caught in a steel trap close by . . . The others had made great efforts to set it free by overturning large stones from the cache in which it had been caught, and they had scratched at the frozen ground around the stone to which the chain was made fast. The trapped wolf could not have done it, for it was caught by the forelegs . . . Next morning their tracks told that they had been to the place where the other wolf had been trapped, and also within ten metres of our sledge camp.'

Wolves have strong territorial instincts, using runways or circuits referred to as 'hunting routes'. These travel ways give access to the territory of a pair or family, generally running through open country. They may consist partly of game trails, cattle or sheep tracks, old wood roads, or even highways in thinly settled areas, and they traverse dry washes or canyons, across low watershed divides or swamps. In cold countries, frozen lakes may become part of a winter runway. Sometimes wolves use runways covering a circuit of considerably more than 100 miles. The width of the run may vary from only a few feet up to a mile or more in country where the animals are hunting. Along the runway there are also high vantage points used for observation; wolves also use these sites for playing or resting.

On these runways wolves have what are commonly called 'scent posts', that is places where they come to urinate or defecate. They are found on or near the bases of tufts of grasses, on bushes, or on an old weathered carcass. These scent posts may be recognized by the scratches made on the ground nearby after the wolf has relieved itself. This habit of having scent posts and a scratch near them is shared with dogs. The scratching of the wolf after defecating or urinating is possibly the vestige of a former habit of burying the dung or urine. As wolves pass over their runways, they stop at these posts, invariably voiding urine and often faeces as well. Wolves are insatiably curious about the presence of alien wolf scents in their territory and this characteristic is made use of by trappers, who may sprinkle wolf urine above the place where their traps are buried. A passing wolf will be drawn to investigate the strange scent and thus spring the trap.

Many instances are quoted of matings between wolves and dogs, both in the case of domestic bitches being covered by dog wolves and of wolf bitches which have attracted domestic dogs to them when in season. It is believed that nomadic Indians purposely crossed dogs with wolves to obtain large dogs for use as draught or pack animals.

The most illuminating account of the life of wolves in the Canadian Arctic is that given by Farley Mowat [46] in his book *Never Cry Wolf*. The author joined the Canadian Wildlife Service and received as his first assignment the task of camping in the deserted Barren Lands: where apart from the wolves there existed only nomadic Eskimo trappers. Mowat was to study the ways of life of the wolves and report on the inroads made by them into caribou stocks.

The author set up his camp in the 'territory' of a family of wolves consisting of father, whom he named George, mother who got called Angeline, four cubs, and a bachelor male who received the name of Uncle Albert. Far from being resented or attacked, the young scientist was irked because the wolves, with whom he would have liked to be friendly, completely ignored him. Mother stayed at home with the cubs, but George and Albert would go out regularly at night to hunt, bringing some food back to the family but secreting the rest in a cache near by, where the mother could go and collect it when it was needed. The family lived in a den, from which they all came out to play and could easily be observed through a telescope.

The family's territory was respected by all the other wolf families around and was regularly marked out by the male wolves by anointing the boundary marks with their urine. So hurt was the author by the wolves' neglect of him that he decided to mark out his own territory and did so, deliberately making his boundary cross that of the wolves. He marked out at about 15 yard intervals some three acres surrounding his tent, a feat which took the whole night and required the consumption of large quantities of tea. When George returned from his hunting, he quickly detected the alien human scents. After some minutes of indecision, he decided to act in a generous manner and made new marks of his own on the external side of the author's boundary, thus conceding his claim to his small territory.

When the caribou migrated north with the melting of the snow, the wolves did not follow them but fed – extremely well – mainly on mice and lemmings. They also fed on various kinds of fish and were adept at driving pike up the streams until they were cornered in narrow bottlenecks at the stream's head.

Meanwhile, Mowat had become friendly with two Eskimos and had learned to communicate with them. One of them was a 'shaman' and belonged to the 'wolf' clan. He claimed to have an affinity with wolves, could understand their language and could communicate with them. As a young lad, his father had placed him for some twenty-four hours in the den with a wolf family, where he had played with the cubs and been fondled by the parents. One day, he heard the wolf message from afar indicating that the caribou were returning and were to be found on the shores of a certain lake to the north. When this message came through, not only George and Albert but the Eskimos also departed to the lake, where they duly found the caribou. On another occasion, the shaman heard news on lupine telegraph that some Eskimo friends of his were coming from the north to meet him; he accordingly went out to find them and duly discovered them where the wolves said they would be.

When one of the Eskimo's husky bitches came on heat, she was introduced to the sex-starved bachelor Uncle Albert, with whom she was allowed to associate in the wild until the end of her heat period. She then duly returned presumably to mother wolf/husky pups, though the end of the story is not told.

These observations reveal something of the highly organized social life and hunting activities of the wild northern wolves. The Eskimos maintained that the wolves' system of language and vocal communication was the equal of the Eskimos' own and that they could readily transmit information between the families of wolves living in adjoining territories, and that messages were sent over large areas, being relayed from one territory to the next.

One last anecdote told by the shaman is worth repeating. It happened that a white hunter visited the territory and shot one of the female wolves, which was suckling four cubs. The male wolf was greatly upset and so were the Eskimos who observed him to see what he would do. He was joined by a dog wolf from a neighbouring territory and together they dug into the dead wolf's den. They each then removed one cub and carried it to the second wolf's den, returning for the other two cubs which were also taken there. The second bitch then reared all the cubs together with her own as one family, a total of ten cubs.

The Eskimos stoutly denied that the wolves made serious depredations on the caribou, saying that they only took the weak and the sick, thus improving the caribou stock and in fact doing them a service. The inroads of the Eskimo trappers, who took fine young animals, many of them cows, appear to have been much more serious. Nevertheless, wolf and man lived – as surely in Palaeolithic times – in perfect amity and respect for each other, and the means of livelihood which they shared was adequate for each.

This author also confirms the organized hunting activity of the wolves when after caribou, both in driving them into traps and in setting up ambushes. He also deals with the control of the population of wolves in an area. The wolves, male or female, remain celibate until they can acquire a 'territory', and may have to wait a considerable time for this. In the meanwhile, like Uncle Albert, they assist other wolves in rearing and feeding their families. In addition, at times when food is scarce – as in the population 'crash' years of the lemmings – by natural means fewer cubs are born and litters are smaller. The mechanism of this has been described by one of us (R.N.F.) in his book *Man, Nature and Disease* [23].

It is tempting to continue these fascinating quotations, which throw so much insight into the hereditary instincts and habits of the wild animals from which our breeds of domestic dogs have been derived. Enough has been said, however, to reveal some of the innate urges which dictate the behaviour of domestic dogs and which, under selective breeding, have come to endow different breeds with different characteristics useful to their human masters. We have already mentioned that litters of wild cubs vary greatly not only in colour and size, but also in temperament. Some cubs are very wild and it is impossible to train them; others are docile and friendly and become faithful and affectionate servants and companions. It is this variability which has made it so easy to breed domestic dogs with such diverse characteristics.

We must now consider in a little more detail the people who first domesticated and selectively bred a wild animal, the conditions in which they lived at the time, and the urges which impelled them to take this far-reaching step.

It is doubtful whether any historian has pondered at length on the deep significance to modern history of that period which is known as the Mesolithic. This period varied greatly in length in different parts of the world, lasting only a few hundred years in some Mediterranean areas, but some thousands of years in more northerly regions.

During the preceding Upper Palaeolithic, which occupied the last glacial period and lasted for many thousands of years, man had adapted himself to his circumstances and was able to provide himself with all the necessities of life as he then understood it. In this time of stability, the different races of man diverged to give the variable characteristics of skin colour, hair and physique generally which still differentiate them.

The ice fringes spread across Europe and Asia, passing over the Alps and Himalayas and in Britain descending as far south as the river Thames. During the summer the ice receded, leaving a lush rich tundra of vegetation supporting vast herds of herbivorous animals which migrated seasonally with the ebb and flow of the ice barrier, attended and pursued by predators.

It is supposed that the Australian aborigines must have crossed into Australia before the end of this period, because at that time the seas were depleted of water since vast amounts were locked up in the arctic ice. Because sea levels were considerably lower than today, man's passage to Australia by way of the Indonesian archipelago would have been comparatively easy. With these people went the dingo, a semi-domesticated form of the Asian wolf, which became feral in the Australian continent.

It is known, therefore, that in Palaeolithic times, one human race had already established community with a wolf form ancestral to one of the basic types of domestic dog. Probably he was content to rear wild-born puppies rather than to bring the animals into full domestication and adapt them to his needs by selective breeding. These needs appear to have been simply for an animal to assist him in his hunting, although members of the family no doubt also lavished some affection on the puppies as they do today.

Other tribes in a similar stage of development passed from Siberia by way of the land bridge, then present at the Behring Straits, into the American continent. Whether they took dogs with them – dogs derived from the northern wolf – is unknown. It is known, however, that wherever these Mongoloid tribes (ancestral to the so-called Red Indians) went they domesticated whatever wild wolf or wild dog was present in the vicinity. In the northern regions, timber wolves were domesticated and from them various local breeds were developed by people of both Indian and Eskimo stock. Typical of these is the rather special breed found around the Great Bear Lake and known as the hare Indian dog, which was only possessed by Indians living around the lake and on the banks of the Mackenzie River. These dogs are affectionate but wild and resemble collies. Those tribes that penetrated further south domesticated the prairie wolf, better known as the coyote, from which one breed of American dog is derived.

In South America there exist a number of forms of forest-living wild dogs from which sprang domesticated dogs. These are ancestral to the various Azara dogs, which are found in many countries of the South American continent and even in the Falkland islands. Another form whose puppies are captured and reared in captivity is the strange-looking small bush dog *Icticyon* (*Speothos*) *venaticus* of Brazil.

Races of man at an Upper Palaeolithic cultural level or its equivalent thus possessed an urge to provide themselves with some form or other of domesticated or semi-domesticated dog. This is strong evidence for the proposal made in this work that our modern breeds have been derived in the main from at least four different ancestral stocks. We may further suppose that with such primitive races of men, having so few possessions, dogs were not kept for sentimental reasons but for the part which they played in various forms of man's domestic economy.

When the end of the Ice Age ushered in the transitional period of Mesolithic culture, some tribes migrated north with the receding ice barrier; they continued to live as hunters, preying on elk, reindeer, polar bears and seals, and catching fish. These people were the ancestors of the Eskimos and of the tribes of northern Siberia which inhabit the shores of the Arctic Ocean.

Other races were compelled to adapt themselves to new conditions in order to survive. These conditions can have been little to the liking of nomadic hunters. The tundra, which had supported the herds of animals on which they lived, was replaced by ever-thickening vegetation, culminating in dark forests of thick-boled trees. The animals of the tundra were replaced by those of the forest, such as the elk and the aurochs, which could not be hunted so easily. Man began to live by collecting edible forest produce, by hunting forest animals and birds, by harvesting sea foods and catching fish. These men were no longer nomadic; their settlements are found between forest and sea or lake, marked by vast middens containing their refuse (including quantities of oyster and mussel shells) and revealing the kinds of provender on which they fed.

In these circumstances, the dog became important to survival; and it is evident that at this stage of human development wolves were not only taken successively into captivity from the wild, but that the age of selective breeding had started. The skill of the northern wolves in tracking by scent would be invaluable for hunting under forest conditions, and also in selecting animals in the trapping and domesticating of ruminants. The second animal to be domesticated after the dog in northern countries was the reindeer. As stated previously, the Lapps of northern Finland are powerless to control their reindeer without the aid of dogs, and it is to be supposed that the reindeer could not have been domesticated unless dogs were available to control the herds.

A lesser but still important function of dogs at this stage must have been, as it is today in cities of the Middle East, to keep the settlements clean by scavenging. Nomadic peoples would have little understanding of hygiene and their settlements could easily become hotbeds of disease unless dogs cleared the refuse.

The life of Mesolithic man was unenviable. There was little comfort in the dark, northern climate and against the ever-encroaching forests he was powerless, having no adequate tools with which to fell the trees. In Mediterranean countries, the Mesolithic period was short, for knowledge of agriculture quickly spread; cattle, sheep and goats were domesticated and the people came to live in comfortable agricultural settlements supported by a shifting but satisfactory form of farming, based on wheat and barley crops. The spread of the Neolithic way of life was accomplished between 8000–2500 BC, starting in the eastern Mediterranean (Sonia Cole [12]). For northern communities the important event which ushered in this change in their way of life was the invention of the polished stone axe, by which the forest trees could be felled and the soil thereafter prepared for agriculture. At the same time, wood from the felled trees could be worked and made into new implements; and even more important, tree trunks could be hollowed with polished stone adzes to make dug-out canoes. Travel by sea and river thus became possible, especially when more sophisticated types of boats came to be built. By these means, in Neolithic times around 3000 BC an era of relatively great prosperity was introduced. The hardy northern peoples, who had entered late into the Neolithic revolution, became enthusiastic traders and were responsible for a large part of the maritime commerce of the time. The well-known 'amber route' across Europe was developed for trading purposes. This route used boat traffic, which passed along the great rivers, chiefly the Rhine and Danube, to bring amber and other northern products to the wealthy Mediterranean peoples.

From the very first, the dog was plainly of great importance to man in many of his activities and was domesticated and trained for the work he was required to do. This took place in several centres, from whatever ancestral stocks were available; although anatomically similar, these stocks had many different characteristics, making the progeny suitable for varying tasks. During the Neolithic revolution, the various breeds were dispersed along man's migration routes throughout the known world. Local needs and preferences determined the pattern of distribution and deliberate selection would result in the production of local breeds, each with special characteristics, each adapted for certain tasks required by the owners.

At this time also came the demand for smaller dogs, probably because by then people were living in permanent settlements and needed smaller dogs for household purposes. In the lake settlements of Switzerland and elsewhere, quite small dogs of spitz types were kept and their remains are commonly found at such sites. From very early times in Tibet and China it was customary to keep both guard dogs and house dogs, a custom which persists to this day. Guard dogs were big and fierce and were kept outside, but the house dogs were small pet varieties.

Pet or toy breeds had been developed before dynastic times in Egypt and the favourite was the so-called Maltese dog of spitz type. Evidently these had been brought from northern countries, probably along the amber route; they were small dogs with silky white coats and have persisted in very similar form for 5,000 years.

Dogs of spitz type were thus introduced from the northern region, where they were developed as early as Neolithic times. For herding sheep and cattle the larger breeds were, and still are, pre-eminent; they spread southwards at least as far as the Rhine to form the main breeds of sheep-dogs.

Throughout the Mediterranean countries, in spite of some influence from dogs of spitz type, the main flow of imported breeds appears to have been from the south and east. Greyhounds, originating in the desert areas of Africa and Arabia, spread eastwards and northwards to form the main canine breeds of the Greeks and the Celts. The very dissimilar mastiff breeds descended mainly from the Tibetan wolf (*Canis lupus chanco*), spread westwards into Anatolia and the Mediterranean world and eastwards into China. This adaptable breed thus produced the main canine populations of classical times. The Hyrcanian and Molossian dogs were of this type; they were greatly prized for their prowess in war and hunting and as fierce guard dogs, both for households and for animal flocks. Dogs of this type were much valued by the Germanic tribes. Thus their distribution came to overlap that of the northern spitzes, and the main areas where they were found came to lie to the north of the greyhounds.

These big mastiff dogs were kept especially throughout the east-west mountain ranges, stretching in an almost unbroken line from the mountains of Tibet, the Himalayas, the Anatolian Mountains, the Alps, the Massif Central of France to the Pyrenees. They are exemplified by the Pyrenean mountain dogs and large dogs of the great dane and bloodhound type. There was also a smaller pointer-type dog, found throughout these areas, a prototype of the retriever/spaniel group. Such dogs have been widely spread for many centuries past and indeed accompanied the peoples involved in the Neolithic migrations. The dog found in the Windmill Hill excavations has been described as a kind of foxhound, but our studies of this skeleton suggest to us that this dog is no other than the prototype form just mentioned.

With their many valuable properties, dogs of mastiff type do not appear to have the herding abilities of the northern spitzes. In eastern Mediterranean areas and even in the mountainous regions of Greece, there was a greater demand for dogs to guard animals than for dogs to round them up and herd them. In certain areas, such as the mountains of the Caucasus and the Hungarian plain, dogs were required which had both herding abilities and the size and fierceness to drive off robbers, wolves and other predatory animals. These needs appear to have been supplied by crossing dogs of mastiff type with spitz shepherd dogs to produce the shaggy, cold-resisting, fierce dogs of the old English sheepdog type; such dogs are typically represented by the Hungarian komondor and Russian owtchar breeds. They are distributed in a wide belt from eastern Russia to Great Britain. The Hungarian and Russian breeds are among the largest known dogs and are of great ferocity, strength and hardihood. The shaggy

25

sheepdogs are afraid of nothing and, while slower to work herds of sheep and cattle, are nevertheless capable herd dogs.

Of the dingo group, little can be said. The ancestral *Canis lupus pallipes* is distributed throughout the Middle and Far East from Syria to India. The dingo is little altered from the ancestral and virtually wild form that was taken by the aborigines to the Australian continent.

Various breeds exist throughout the Indonesian Archipelago and New Guinea. Towards the west, the type changes slightly in conformation and carriage of tail to form the pariah dogs. Whether these are indigenous in the areas where they are found or were introduced in early times, whether they are descended from fully domesticated stock or have always existed in semi-domesticated form cannot be said. Some have been fully domesticated both in Asia and in Africa, as for instance the basenji and other hunting breeds found in more southerly parts of the African continent. These dogs may have spontaneously adopted the life of scavengers near human settlements and may never have been fully domesticated or bred for useful purposes by man; if so, they are in a self-adopted feral state, just as were wild dogs in Mesolithic times. Their northward spread and influence on the northerly breeds will be studied in chapter 6.

In ancient Egypt from very early times at least five different breeds of dog were known and kept for different purposes. These included dogs of basenji type, of greyhound type, of Maltese type, of mastiff type, and a spitz type resembling the chow. Possibly the latter indicates the existence of trade routes with China, along which were brought dogs developed in the orient. Among the many breeds of dogs living today, there are few whose prototypes were not already in existence even before the days of the First Dynasty in Egypt. The domestic dog has shown little capacity to vary in its essential characteristics beyond those which had been developed at that time.

Breeds of dogs of comparable variety were developed in the Central and South American empires of the Aztecs and Incas, though the picture was complicated because of the importation to Central America and Peru of dogs of spaniel type by the conquering Spaniards. The smallest dog in the world, the chihuahua, was developed from the Central American breed crossed with a dog of Manchester terrier type. The small toy breeds of the Aztecs and Incas appear to have been developed largely as temple and sacrificial animals. Further breeds of dogs were, as already noted, developed from the wild canine races of South America.

Chapter 3

Dogs in Ancient and Classical Times

THE classification of dogs of Mesolithic and Neolithic times has been discussed by Zeuner in his *A History of Domesticated Animals*[70]. The various breeds of dogs, or supposed breeds, have been given subspecific names which appear to have little validity, since the relationship of one form with another has not been established and the same form may well have two different subspecific names. There is a northern dog, *Canis familiaris inastronzewi*, which may be ancestral to or related to various forms of northern dogs. There is *C. f. palustris*, the small dog of the lake settlements, also probably of spitz type; some skulls of this dog were a mere four inches in length. Another dog, *C. f. poutiatini*, found near Moscow, was ascribed to Neolithic times; its skeleton is exceptionally well preserved and is said to show characters resembling those of the dingos. Very small dogs, *C. f. spolleti*, have been recovered from lake dwellings in Bosnia, Italy, the Ardennes, Switzerland and Austria. An early sheepdog type known as *C. f. matris-optimae* has been found at various sites in the Rhine valley.

These data are important since they show that the basic breeds of dog were present during Neolithic times and that they had achieved widespread dispersal. The Neolithic revolutions started in Middle Eastern countries around 8000 BC and reached Britain and Scandinavia by 2500 BC. Avenues of trade had been opened, as described in the last chapter; both people and animals had become more widely dispersed and cultures had intermingled. The spread of canine breeds over a wide area was thus to be expected. The new stone culture was based on clearing of agricultural holdings by stone implements. Preparation of the land was inefficient and the crops obtained were poor in yield and quality. The Neolithic, like the Mesolithic, must be regarded as largely transitional, although during this period man did provide himself with some degree of comfort and prosperity.

The introduction of metals emancipated man to a more or less civilized way of life. It became possible to cultivate land properly by hoeing or digging deep, and primitive ploughs drawn by oxen were introduced. Towns and cities were established, but the era of the 'city states' was not altogether a happy one. Cities needed to be fortified and defended and the possession of riches brought avarice.

Hitherto, the uses of dogs had been simple: to draw sledges, to assist man in his hunting forays, to provide food and skins for clothing, to guard homesteads and flocks. The newer, more settled way of life brought more varied demands for dogs of different types and allowed greater skill to be exercised in perfecting breeds for different purposes. Hunting, instead of being a necessity of life, became a sport and a sport of kings at that. The quarry was not small hares and rabbits and foxes hunted today, but wild cattle, horses, tigers, and other large animals. For these purposes very large dogs of mastiff type were bred; these were scent-hunting hounds and their pictures can be seen on bas-reliefs and tablets from Assyria and Babylon. In Egypt and Arabia, on the other hand, the sight-hunting grey-hounds were developed into breeds of different sizes suitable for coursing animals varying in size from hares to gazelles. Greyhounds became especially popular in lowland areas of Greece. They were also adopted by the Celts and during their migrations, greyhounds became widely distributed in the areas settled by these tribes. The Germanic tribes of the north, who lived in more forested areas, owned dogs predominantly of mastiff type, but their domain was also the meeting point of northern spitz and mastiff.

In those days, dogs of large size were required for protection of the herds and flocks in mountain areas as well as for hunting. Very large dogs of mastiff type were kept by wild mountain tribes, probably to protect themselves and their herds as much from brigands as from wolves and other wild animals. Some of these we have already encountered in the Molossian and Hyrcanian dogs and in the Pyrenean mountain dogs which were ancestral to many of our modern breeds. Thus, while the popular dog of ancient Greece was the greyhound, in the more mountainous areas large, fierce dogs of mastiff type were kept.

These very large dogs were required for yet another purpose – as shock troops in battle. The Celts, during their invasion of Greece, used dogs of greyhound type and it is believed that the Persians also used mastiffs in war. The method seems to have been to unloose some hundreds of these dogs before the advancing infantry; they were trained to hurl themselves upon the enemy and must have been often impaled upon his spears. The

28

defending troops, thus demoralized and their spears out of action, would easily be demolished by the attacking force.

To have sufficient numbers of very large and fierce dogs for this purpose would present no difficulty, because of the rapid rate at which dogs breed and because of their early maturity. Their maintenance, however, would be extremely expensive. According to Herodotus, for instance, Indian dogs were highly valued by the Persians and the revenues derived from the taxation of four large villages in Babylonia were devoted solely to the upkeep of the Indian hounds belonging to the Persian satrap. As an example of the numbers of dogs that were kept by monarchs in ancient times, it is recorded that among the animals exhibited at Alexandria during an all-day parade by Ptolemy II were 2,400 hounds of Indian, Hyrcanian, Molossian, and other breeds. These were predominantly of mastiff type. The very large dogs required for hunting large game, for protection and for war, have few counterparts today and are far less widespread.

Peoples of ancient times also required dogs of smaller size for various purposes. Although little mentioned in literature, it is to be supposed that a form similar to the Windmill Hill dog was widely distributed throughout the mountainous regions of the world. That this was so is largely conjectural and rests on the fact that such dogs are still distributed throughout these regions and certainly were so at the beginning of the Christian era. This is a basic breed from which our sporting dogs were derived, particularly the spaniels, setters, pointers, dalmatians and retrievers.

Today such dogs are found in the possession of people from the Caucasus throughout the Alps, Vosges, Jura, and Massif Central to the Pyrenees. They have different breed names – or none – and are represented by the Breton spaniels, the French braques and dogs in the Caucasus similar to the one figured in plate 1. The dalmatian is of this type and historically it is known to have been brought by the gypsies from India, so that its distribution must have extended far to the east. These are dogs of the common man – his companion and his watch. They were not carved on tombs or described in ancient literature; nevertheless they were the backbone of a widely distributed canine population derived from very early times. These dogs do not have shortened muzzles and they have floppy ears; they are very variable in colour, markings and coat type, though mostly the hair is silky. They are probably derived chiefly from mastiff origins.

The development of cities and civilized ways of life resulted in the acceptance of dogs of suitable breeds as inmates of the household and as family pets. During the whole of the classical period, from early dynastic Egyptian, Cretan and Greek to Roman times, the small Maltese dog, which is still popular today, was in great demand as a 'toy'. Dogs of pomeranian type also began to make their appearance and other miniature breeds were being developed as far away as Tibet and China. Thus a whole range of dogs of various breeds suitable for different purposes came into being and they came to be regarded with affection. Epitaphs were written about them when they died and they were given the same burial rites as their masters. Shepherd dogs, the common guard dogs, as well as war dogs, were equipped with spiked collars to give them protection from wolves and other attackers; household dogs were provided with simpler collars and leather straps for a leash.

Most information on breeds can be gleaned from representations in stone carvings or on tablets. No writer of classical times has given descriptions from which any real information about the appearance of the various breeds can be obtained and their names were almost universally taken from the place or area from which they came. We shall, however, review what is known of the dog populations of different countries in ancient and classical times and see what can be discovered.

Dogs in Biblical and Middle Eastern Countries

Throughout Biblical times, dogs of pariah type appear to have infested eastern cities, though possibly not those of Egypt. These animals were certainly derived from dingo-like ancestors, descendants of the pale-footed Asian wolf. It is not known whether they were ever domesticated deliberately and became secondarily feral, or whether they were the descendants of wolves which found an easy livelihood by scavenging around human settlements. Nor is it known whether at any time the pariah dogs, as their mode of life might suggest, crossed their blood with that of jackals. It is evident that by Biblical times they were already an unclean pest which people feared to touch. Whether, as seems possible, there was some rational medical reason for this in early days is also unknown. It is possible that they may have been widely infected with the minute dog tapeworm known as *Echinococcus granulosus*, the larval form of which causes the condition of hydatid cyst. The larvae of this tapeworm can infect human beings, sheep, donkeys, and other herbivorous animals; they multiply greatly within a cyst capsule which grows to an enormous size, sometimes in the liver and other vital organs thus causing death. In recent times, hydatid has been a major scourge around the sheep stations in Australia and New Zealand.

Although in Biblical times pariah dogs were held in abomination, other breeds were favoured, such as Maltese dogs among the Jews and salukis among the Arabs. It must remain a mystery as to why the unfortunate pariahs remained so ill-regarded and were not taken into proper domestication as were other breeds. The dingos have been good hunting dogs to the Australian aborigines and have produced strains in many Far Eastern countries which are favoured by the natives. Basenji-type breeds were apparently present in Egypt in early times and were popular as household pets. It would seem also that canine breeds descended from the pale-footed Asian wolf have been crossed with other breeds as far to the north as the territory of the Lapland spitzes.

In spite of their detestation of dogs, the Hebrews used them as watch-dogs for their houses (Isaiah 56–10) and for guarding their flocks (Job 31). They appear, however, to have been mostly uncontrolled, and troops of hungry and semi-wild dogs used to wander about the fields and streets of the cities devouring dead bodies and other offal.[1] They became so disliked that fierce and cruel enemies are poetically styled 'dogs' (Psalms 22, 16[11] and 20[12]); they are regarded as an unclean animal (Isaiah 66, 3[13]); and the terms 'dog', 'dead dog' and 'dog's head' were used as epithets of reproach or humility in speaking of oneself.[2] To this day, descendants of the dogs supposed to have devoured all of Jezebel except for the palms of her hands and the soles of her feet prowl around the walls of Jezreel in search of offal and carrion.

It is not clear whether shepherds' dogs were the same as pariahs; they are supposed to have been very fierce and were possibly a different breed. In Rabbinical literature two types of dogs are recognized, the one being classed as 'domestic', the other as a wild animal. The former was accepted and to cross the two was forbidden. Obviously in very early times, domesticated dogs were useful: Abel himself was a shepherd and no doubt had his shepherd dogs. Thus eastern races distinguish between the animals they had captured and bred for their needs and those which came to live around their settlements, which were dangerous and unsavoury. To this day the Arabs sharply distinguish pariahs from their own domesticated breeds.

---

1 Reference is made to them in this sense in Kings 1–14[1], 11; 16, 4[8]; 21, 19[8] and 23[4]; 22, 38[5]; 2 Kings 9, 10[6], 36[7]; Jeremiah 15, 3[8]; Psalms 59, 6[9], 14[10].

2 1 Samuel 24, 14[14]; 2 Samuel 3, 8; 15, 16, 9[16]; 2 Kings 8, 13[17].

The dogs of Babylonia and Assyria

Two types of large dogs kept by the Babylonians and Assyrians were carved in mud and stone; one is clearly a mastiff, the other a greyhound. In addition, these peoples had a water dog of spaniel type, an earth dog for burrowing and no doubt other breeds as well for special tasks. Babylonians used dogs for a variety of purposes, including for hunting lions; for driving sheep; as watchdogs (these were pariahs); as household pets; and for purposes of magic and medicine. Unwanted children were also thrown into pits as prey for dogs.

In the Middle East, dogs were often buried in the kings' tombs; for instance in tombs of the Hittite capital of Hattusas (2400–2200 BC) dog skeletons were found with ritual deposits of sacrificed cattle.

The kingdom of Assyria flourished in the 2nd and 1st millennia BC. The four best known rulers were Sargon (722–705 BC), his son Sennacherib (705–681), his grandson Esarhaddon (681–669), and his great grandson Ashur-banipal (669–626). These were the great years of conquest, when Babylon was subjected and the Assyrian empire reached the Mediterranean and extended even into northern Egypt. The fine reliefs now in the British Museum showing the king hunting lions from his chariot, accompanied by his gigantic mastiffs, were recovered from Ashur-banipal's palace at Nineveh. Undoubtedly, these show a method of hunting with hounds which had been in use for many generations, during which the hounds were especially bred for the purpose (plate 2).

Ashur-banipal was the last of the Assyrian emperors. After his death Nineveh fell, the empire collapsed and all his territories in Asia Minor were absorbed into the Persian empire. The hounds, however, evidently continued to be bred both in Persia and Macedonia and were believed to be ancestral to the Molossians of Epirus, where they were introduced during the Graeco-Persian wars.

Dogs of ancient Egypt

In predynastic Egypt, as elsewhere, dogs were kept by Neolithic peoples and different breeds had already been acquired. Civilization started early in Egypt, crops being grown on the new mud flats left after the Nile inundation. The approach of the flood was heralded by the appearance of the star Sirius, when the people hastened to remove their flocks to higher ground and abandoned the lower pastures. Sirius came to be regarded as their god and protector and, because of its watchfulness and fidelity, it was likened to the dog and worshipped as the 'dog star'. To the Egyptians at any rate the dog star was an object of veneration, though in later times it came to be associated by other peoples with the onset of summer heat, the so-called 'dog-days'. Thus in hot countries it was deprecated rather than venerated and the name Sirius itself is derived from the Greek word meaning 'scorching'. Sirius was the dog of Orion the hunter and with him pursued the rain stars, the Pleiades.

In ancient Egypt almost all animals came to be worshipped at some centre or another and the dog was no exception. Apparently the inhabitants of the town of Cynopolis particularly venerated the dog, though where this town was cannot be ascertained. Possibly the whole story is a figment of Herodotus' fertile imagination. According to him, the wolf was worshipped at another centre named Lycopolis, a Greek term meaning 'the city of the wolf', just as Cynopolis means 'the city of the dog'. In the same way the names of some villages in sub-Saharan Africa are associated with animals; for instance in the district of Lango in Uganda, various disparaging names are attached – presumably by strangers – to the villages; one such village is called Ngotokwe, which being interpreted means 'the place where they fornicate with jackals'.

Dogs were held in veneration over the greater part of Egypt, if not

Figure 1. Breeds of Egyptian dogs, from the tombs at Beni Hassan, 2200–2000 BC. (After Ash.)

universally worshipped. When they died, their bodies were properly prepared by the embalmers; they were wrapped in linen and deposited in tombs at public expense, while bystanders beat their chests in token of grief and uttered lamentations in the dog's honour. In every town a grave-yard was devoted entirely to dog burials.

Among the mummified dogs are small animals, probably of basenji type, which wore collars made of leather or metal moulded to represent leaves. There are no records of pariahs, but it is certain that dingo-type dogs had reached ancient Egypt at a very early date, for they are shown by repre-sentations of animals remarkably like the basenji breed of today.

The ancient Egyptians domesticated, or attempted to domesticate, almost every known living beast on which they could lay their hands. Pariah dogs may have been crossed with jackals; this would present no special problem for the ancient Egyptians because of the unique place in their pantheon occupied by the jackal itself. The jackal-headed god Anubis was the god of death and associated with the funeral rites which played so great a part in their religious thinking.

Cats, even more than dogs, played a part in the religions of ancient Egypt. It is also evident that they were inmates of the ordinary Egyptian household, as seen in the intimate family scenes depicted, not on the tombs of the pharaohs, but on those of the queens, the nobles and princes. There are representations of the family at breakfast, with the cat under the table eating a fish. In the same way, dogs were evidently much loved and, apart from their uses in the hunting field and in war, they were family pets and guardians of property. These dogs were given names, just as we give names to our dogs; some that have been deciphered are 'Salekai', 'Xabesu', and 'Akena'.

The first domestic dogs known in Egypt have been described as of spitz type; but they are more likely to be of dingo/pariah type, with strong characteristics of the basenji. These are the 'Khufu' dogs, depicted in a tomb of the period of the Pharaoh Khufu (Cheops) of the Fourth Dynasty, around 2700 BC. No dog of any kind was depicted in ancient Egypt before this date. However, in tombs of this dynasty, house dogs are shown attached to the chair of the master of the household. They were named 'thesam', meaning hound or ordinary dog. Dogs are also shown in tomb pictures of the Old Empire attacking antelopes and even lions; these were more of the mastiff type.

In all, the ancient Egyptians seem to have kept some six different breeds of dogs, as well as mongrels derived by crossing them. These were: (1) a small terrier-like form (Maltese); (2) a dog of pointer type (*vide* the

Windmill Hill dog); (3) the Khufu or basenji type; (4) a mastiff type; (5) a dog resembling a mastiff/great dane cross; (6) dogs of both greyhound and saluki types; the latter were probably derived from Arabia. Even in Upper Egypt, dogs of greyhound or saluki type were present during the second half of the fourth millennium, as shown by paintings on Badarian and Amratian pots.

In a paper published in 1875, Birch described a tablet of Pharaoh Antefaa II dating from around 2000 BC. It showed the Egyptian pharaoh with four dogs of different breeds painted in different colours, and is of exceptional interest, since the four breeds kept by the ancient Egyptians are clearly depicted and their uses are distinguished by the type of collar they wore. The first dog was called 'Bahakaa' or 'Mahut' and wore a

Figure 2. The four dogs from the limestone stele of Antefaa. XII Dynasty, c. 2000 BC. (After Ash.)

narrow collar of single width tied in front. These names seem to apply to the breed rather than to the dog itself, because some other hieroglyphs appear to indicate that the dog's name was 'white antelope'. This dog had pendant ears and in some respects resembled the primitive 'pointer' type.

The second dog appears to have been called 'Abakaru', though whether this was was the name of the dog or the breed is not clear. The collar was broad and consisted of four bands tied in front. The dog had a more

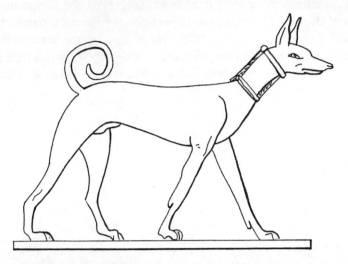

Figure 3. Egyptian greyhound with a restraining collar, presumably untrustworthy. (After Ash.)

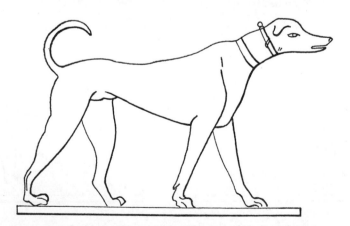

Figure 4. Egyptian greyhound with floppy ears but normal tail. (After Ash.)

marked 'stop'[1] than the first dog. The muzzle was sharply pointed and fox-like, the ears pointed and held erect, with the tail curled close to the left side of the back. It is evidently of the same breed as the Khufu dog depicted on Fourth Dynasty tombs some 1,600 years earlier and is probably a basenji.

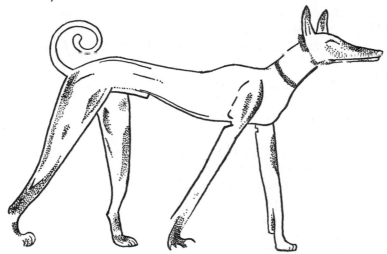

Figure 5. A large greyhound with the curly tail preferred by the Egyptians. (After Zeuner.)

The third dog on the tablet was evidently of mastiff type. Its name was 'Pahates' or 'Kami'; the latter name means black. The collar consisted of two bands tied in front.

The fourth dog was shown between the pharaoh's legs and appears to be of the type of the ancient Egyptian house dogs. Its name was 'Tekai' or 'Tekal', which means either 'laying waste' or 'destruction'. This dog may well be a mongrel with some mastiff blood. It wore a collar of two strands, but the method of fastening it is not shown; the collar seems to have metal studs in it.

It is most probable that many more breeds of dogs were kept in ancient Egypt than those mentioned. There was a hound-type dog, with a low body like a dachshund; one is depicted as black and liver, with pointed nose and erect ears. Sometimes the colour of these dogs is shown as white with brown spots, or yellow and white with red flecks. Another dog sometimes depicted was short and thickset, with spots and a yellow skin.

[1] The 'stop' is the point where the skull joins with the nasal bones. Where the frontal sinuses are large, there is a greater angle at this point. Thus a collie, for instance, has no stop, but a retriever has a normal stop.

Yet another was thicker in the body, longer in leg and had a curly tail and long pointed muzzle.

It is evident that the Egyptians, as one would expect, developed numerous breeds from the basic types described; it would also be characteristic of them to have developed jackal/dog, wolf/dog, and fox/dog crosses. That such crosses could have had some influence on later breeds has already been suggested and the great variability of some breeds of modern dogs could be ascribed to this cause. Loisel [39] states that for hunting the Egyptians used dogs, cats, wolves, hyaena-like dogs, real hyaenas, leopards or cheetahs, and even lions. Who can say whether alien blood may not have been introduced by these adventurous people, even perhaps including Cape hunting dogs (*Lycaon pictus*) and the wild dog of Abyssinia (*Canis simensis*)?

The earliest dogs of saluki type (greyhound group) are depicted on the tombs at Hierakonpolis of the Twelfth Dynasty. These dogs were derived from desert wolf forms, possibly similar to *C. l. arabs*. This may have been the wolf which the Egyptians used for hunting up to the Twelfth Dynasty but not later, possibly because it had disappeared from Egypt by that date. Among their other crosses, the Egyptians crossed dogs of mastiff type with their salukis or greyhounds, thus producing a breed resembling the great dane. Whether the ancestral great dane breeds were originally developed in Egypt is unknown, but this would appear to be improbable.

Dogs of Greece and Rome

Rome was traditionally founded in the year 753 BC, a date which archaeologically would appear to be reasonably accurate. The city was founded by the legendary Romulus, son of Mars by Rhea Silvia, daughter of Numitor king of Alba Longa. Romulus with his brother Remus was left on the mountains by his great uncle Amulius, who had usurped the throne. The twins were suckled by a wolf and brought up by Faustulus and his wife Acca Laurentia. Remus was slain by Romulus because of a quarrel over the site of Rome, and eventually in 716 BC Romulus was carried to heaven in a chariot by Mars and was worshipped as the god Quirinus.

If it teaches nothing else, this strange story with its circumstantial evidence shows that wolves existed in the Latin peninsula around that time and that the relationships of these animals with man were sufficiently close as to make it credible that human children could be suckled by a she-wolf. The people who founded Rome, in spite of the oriental influence of the Etruscan empire, were primarily Indo-European and came to Latium at about the same time (1000 BC) as the Dorian invasions of ancient Greece. They were Neolithic agriculturists and pastoralists, probably very similar to those who spread to the west during the Neolithic revolution.

Throughout both Greek and Roman history, it is evident that the backbone of the canine populations was the mountain shepherd dogs, used both for herding and for protection of the flocks. We have already seen that these were predominantly of mastiff type, although the offshoot pre-spaniel types were widely distributed through mountain areas from India to Britain. Just as the terriers of Britain in more modern times have developed into numerous local breeds, so these dogs of Greece and Rome developed and were known, as are the terriers, by the names of localities.

Whereas the country people clung to their traditional breeds, perhaps sometimes crossing them with spitz dogs from the north so as to acquire greater herding skill, the townspeople and the country landowners developed other breeds more suited to their sporting and other requirements. For instance in Greece hunting was the sport of the gentry who developed the greyhound, originally acquired from North Africa, as the most popular dog. It is this dog which is depicted on Graecian urns and amphorae. The Roman hunting dogs, needed for more mountainous regions, were developed mostly from mastiff breeds acquired from the length and breadth of the Roman empire; thus the merits of British and Gallic and other breeds were hotly argued and no doubt a great many crossbreeds were produced. Middle class and wealthy people, who dwelt mainly in cities, wanted smaller dogs for house protection, as pets, and as 'toy' dogs popular with ladies. Similar developments in Tibetan and Chinese dogs will be noted in the next chapter.

Among the sophisticated Greeks and Romans, the art of hawking and netting birds was unknown until introduced to Rome from Britain. These arts were originally developed in the Middle East, where gazelle were hunted by a method in which both saluki hounds and hawks were used. In Britain, combined hunting by hawk and dog had been introduced possibly by trading Phoenicians and caused a demand for certain specialized types of dog. Apart from dogs of poodle type, such breeds were not developed in Greece and Rome in classical times.

That watchdogs were kept in ancient Greece from very early times is shown by the fascinating story of Odysseus' dog Argus. After ten years of wandering and adventure following the Trojan war, Odysseus reaches his home in tattered rags in the guise of a beggar. After all these years, his wife Penelope is on the point of yielding to one of her many suitors. Neither his wife nor his faithful servant recognized Odysseus; only to his dog Argus is he immediately known. The date of this story according to Homer and now believed to be accurate is about 850 BC.

Among the savage dogs known to the ancient Greeks were Hyrcanian dogs from India, said to have been crossed with tigers on account of their extreme ferocity. Locrian dogs were employed mainly in hunting bears, while Pannonian dogs were used in war as well as in the chase; they made the first charge on the enemy and were perhaps the original 'dogs of war'. The Molossian dogs of Epirus were also trained to war and to fight in the ampitheatre as well as to hunt. Initially they were shepherd dogs, owned by the tribe of the Molossi of Epirus, whence came the mother of Alexander the Great. Apparently these dogs were untrustworthy and were said to have only one redeeming feature, namely great attachment to their owners. This was reciprocated and the Molossi would weep over their dogs when they died or were killed.

Dogs were important in Greek legend because the gates of Hades were guarded by the giant mastiff Cerberus. Only two mortals ever passed him: Orpheus in search of his lost wife Eurydice entered Hades to bring her back to the world, but lost her at the last moment because he could not resist a last look behind; and Hercules who, as one of his labours, successfully fetched the many-headed dog from Hades.

In the various sanctuaries of Asklepios at Epidaurus and elsewhere, the temples were haunted by sacred dogs. A blind boy was allegedly cured when a dog licked his eyelids and another was cured of a tumour in the neck (goitre?) by the same treatment.

Around the year 530 BC, the Greek philosopher Pythagoras paid a visit to Egypt and on his return founded a new sect at Croton in southern Italy, where he taught that when the body died the soul entered into the bodies of animals. If a disciple was dying, he caused a dog to be held to the mouth of the man to receive the parting spirit and he said there was no animal to perpetuate his virtues better than the dog.

Little object could be attained by studying in detail the dogs of Rome at the time of the empire. They were derived from all over the world and must have been a motley crew of different sizes, shapes, and colours. They were used for a multitude of purposes, including food and fighting in the arenas. The Romans themselves more or less abandoned any attempt at classification, lumping various breeds together according to their uses. Thus house dogs were *Canes villatici*; shepherd dogs *Canes pastorales pecuarii*; sporting dogs *Canes venatici*; pugnacious dogs, suitable for war or arena, *Canes pugnaces* or *bellicosi*; dogs which hunt by scent *Canes nares sagaces*; dogs which hunt by sight *Canes pedibus celeres*.

Among the recognized breeds were the following. The Umbrian were used for hunting and shepherding; they were lively and keen with a good nose for scent, but not very courageous. The Sallentine or Calabrian were one of many kinds of sheep dog. Etruscan dogs were apparently something like the modern pomeranian; they were not built for speed, but had a good scent and might be crossed with the Laconian for hunting. Molossian dogs were large and heavily built; they came from Epirus and were esteemed as housedogs and for shepherding. Laconian dogs were lightly built, skilful hunters and also good as shepherd dogs. Cretan hounds were also rather lightly built and were good hunters, but apparently not used as shepherd dogs. Sassalian hounds were similar to the last two, but heavier. The last three breeds appear to have been true hounds and to have hunted by scent.

Figure 6. Maltese dog in Graecian times. (After Keller.)

Acarnanian hounds, on the other hand, hunted silently and by sight, thus being of greyhound type, as were the Vertragi from Gaul. These hounds were famed for their speed and used for coursing hares. British hounds were rough-haired and were used as trackers; possibly they were early terriers, as already suggested.

Finally, there were the pet dogs known as *catelli* or *catellae*; probably most of these were varieties of Maltese dog, so popular for centuries in ancient lands. The emperor Claudius had a white pet dog almost certainly

Figure 7. Small pet dogs in Greece. (After Keller.)

Figure 8. Molossian dog. (After Keller.)

of this race. These Maltese dogs were uncommon in that they were predominantly of spitz type. There is a strong probability that small dogs derived from the Far East may also have been kept; these would have been of the pug type, possibly somewhat resembling toy spaniels.

Chapter 4

Dogs of the Far East and America

THE Far Eastern countries with which we are mainly concerned are India, Tibet, and China. India has already been mentioned as the source of the very fierce Hyrcanian hounds described by Aristotle and Xenophon and alleged to have been crossed every few generations with tigers. These hounds originated in north-west India, probably Baluchistan, and it is more than probable that their savage characteristics were maintained by back-crossing with some wild Canidae such as the pale-footed Asian wolf. It is even possible that they were crossed with the very savage Indian dholes, of which a description will be given in chapter 11.

To this day there exist in India two breeds of dogs which are difficult to classify; these are known as the bunjara and the polygar, both from the northern regions of the subcontinent. The bunjaras stand 24–28 inches at the shoulder and have the reputation of being extremely ferocious. They are dark brown or grey shading into black and have the close coat of a mastiff, but lifted on the elbows and hocks and feathered down the inside of the forelegs like a spaniel. The tail is carried low except when the dog is excited and there is a respectable brush, especially at the tip. The ears are flat to the head and raised only on occasion. They are used exclusively for hunting. The polygar is larger than the bunjara, being 27–31 inches at the shoulder. They are said to be less ferocious, but very plucky; the colour is bluish-grey or dark grey. These dogs, like the hairless Mexican dogs, are almost bare except on the head, which is sleek. They are used for hunting tigers and panthers and are believed to be derived from dogs of bunjara type crossed with greyhounds.

It is not unreasonable to suppose that in these two breeds we see relatives, if not descendants, of the savage Hyrcanian dogs of classical times. Apart from these rather special breeds, India is full of interest as a home of primitive Canidae. It would seem that it was in parts of this sub-continent that the native pale-footed Asian wolf passed into semi-domestication as the dingo. Zeuner found dingo-type pariahs in north-west India and figures a typical little dingo belonging to the pre-agricultural Chenchu tribe. Throughout India and Indonesia there still survive tribes closely resembling the Australian aborigines, and there also exist in these areas pariah dogs of dingo type. The true pariahs of India, that is to say dogs of dingo type with the characteristic curly pariah tails, are to be found in many cities fulfilling their usual role as scavengers. It is said (*vide* Pocock) that these pariahs often mate with jackals and that the strain is consequently mixed. The jackals of northern India, unlike those of anywhere else, sometimes combine to hunt in packs like wolves and are perhaps closer to the common ancestor of these two groups.

It would appear that the bunjara dogs were originally derived from mastiff stock. In addition, true mastiffs are present in northern India, as would be expected in the more mountainous regions. The skull of one such is depicted in plate 36, though the muzzle of this dog is unusually long for a mastiff, possibly indicating admixture with greyhound blood.

Greyhound stock reached India in early days, probably brought by the Greek invaders under Alexander the Great and during the time of the Seleucid Empire which succeeded him. We shall see in a later chapter that the term 'Seleucid' may be associated with the word 'saluki' applied to one of the main branches of the greyhound group of dogs.

Both in Tibet and China the dog populations are also very mixed and it is difficult to unravel the various influences which have contributed to their composition. In classical times the enormous mastiffs first make their appearance with the Assyrians and Babylonians, but most authorities believe that the true mastiff stock originated in Tibet and that the scion was the woolly Tibetan wolf, *Canis lupus chanco*. It is believed that they were taken to the mountains of Anatolia during the westward movements of Mongoloid peoples, probably as shepherd and guard dogs, and that the Assyrian, Molossian and other breeds were developed in those areas. Other authorities hold that the mastiffs were developed in Anatolia itself and that dogs of mastiff stock reached Tibet by eastward migrations. This view seems less probable, first because of the presence in Tibet of the most suitable ancestral wolf, and also because the Tibetan mastiffs do not show the specialization due to selective breeding that is found in western regions. Furthermore, the main migration routes were from east to west and thus a Tibetan origin seems more likely.

At a very early date, Mongoloid people settled the plateaux and valleys of the Tibetan mountains and developed a culture characteristically their own. In ancient times Tibet was the land of legend, the 'roof of the world', a land of jewels and djinns and fabulous wealth. It was also a land of religion and mysticism, with a primitive, intelligent, but ignorant and poverty-stricken peasantry, who supported an elite of cultured but ineffective priests devoting their lives solely to philosophy and religion.

Nevertheless in classical times this country was in all probability less isolated from the western world than it is today. There was free communication with Tibet and China by the caravan route across the Gobi desert. Chinese silks were traded for the enamels and jewellery of Rome and Egypt by means of camel trains, no doubt organized by the wealthy merchants of Baghdad and other Middle Eastern cities. Thus in both Tibet and China, influences on the dog populations may be expected from western sources.

The Tibetan peasants possessed two main types of dogs, one large, the other small. The large ones were of mastiff type, used for guarding flocks and herds and for protecting the home; the small dogs were kept inside the houses as pets, with perhaps symbolic significance as guardians of the home from evil spirits. The mastiffs were said to be of mixed types, some being pure mastiff and others more like chows. It is unlikely that the mastiffs of Tibet are pure-bred as the chow itself is of spitz type, probably mixed with pariah, and was brought from more northern regions by the ancestors of the modern Tibetans.

The small dogs of Tibet were introduced into western countries, where two main breeds are recognized in the 'Lhasa Apso' and the 'Shih Tzu'. There is also a so-called Tibetan terrier, though it is unlikely that this could be a true terrier of the western type.

In contrast with the variability of the dogs kept by the common people, definite breeds were developed in the monasteries, where monks or servants had the leisure to devote to the art of the specialized breeding of dogs. This practice was probably intensified after the introduction of Buddhism into Tibet from India around the year 700 AD. This religious cult with a new leader and a new emphasis gained momentum and many of the enormous monasteries, such as that of Lhasa – the largest building in the world – were founded. It is clear that dogs of pekingese type were kept there, and indeed the figure of one is depicted in the fascinating book of His Holiness the Dalai Lama [18], *My Land and My People*, as the lion of Tibet.

In these monasteries, then, were developed breeds of small dogs similar to the pekingese, and the stimulus to breed specialized lion dogs of symbolic significance probably spread to China from Tibet. Buddhism reached

China about the beginning of the Christian era, but spread in Lamaist form from Tibet after 700 AD and it was at this time that the small lion dogs were developed in the imperial palaces. There seems little doubt that in Roman or even earlier times, Maltese dogs were traded to Tibet and that Maltese blood had entered into the development of the pekingese. This would account also for the conformation of other miniature breeds that became popular in Far Eastern countries, such as the 'Lhasa Apso', the 'Shih Tzu' and the Japanese spaniel. Such an admixture could account also for their beautiful silky hair.

The influence of Tibet on the dogs of China, however, started long before Buddhist times. It was said that in the year 1121 BC the people of western China sent to the Emperor Wu Wang a dog of the 'Ngao' race from Tibet. The western Chinese province of Shantung, which abutted on Tibet, was famous for its dogs of mastiff type and mastiffs were certainly present in China from very early times.

Evidence of the types of dogs present in early China is available from studies of tombs of wealthy people of the Han dynasty, which lasted from 206 BC to around 200 AD. In the days before Confucius (c. 600 BC) great men were buried with their concubines, attendants and favourite animals (as was also the practice in Middle Eastern countries). Together with one such eminent personage were buried alive no less than 66 people and 190 animals including horses, leopards and tigers. At a later date, owing to the influence of Confucius, effigies were substituted for living creatures and among the tomb furniture of the Han era a number of pottery dogs has been discovered. They include watch dogs, hunting dogs and larger ones, but none of pekingese type.

The watch dogs are similar to the ordinary smooth-haired breeds still common in China. The hunting dogs, strangely enough, are of greyhound type and are depicted coursing hares in company with falcons, evidence of cultural contacts with the Middle East. The larger dog of chow type is also mentioned in the ancient *Book of Rites*, where a carcass is depicted hanging from the rafters in a kitchen next to a plucked chicken, a leg of mutton and other joints. The smooth-haired watch dogs are evidently of primitive pointer type, which we have already traced in unbroken sequence throughout the east/west mountain ranges of the northern hemisphere. Dogs of the large mastiff types are not present in these tombs, although as already mentioned they were kept in Shantung province. In Assyria they were used especially for hunting with horse-drawn chariots; no doubt the Chinese did not hunt in this way and so had no need of this type of dog.

47

Apart from the hunting dogs and the chow chows or larder dogs, the common dogs of China were small animals like pugs with short noses, known as 'ha pa' dogs. These little dogs were prized possessions of the people and are still to be found in China, where people value them and are fond of them. The term 'ha pa' means under-table dogs; this is a reference to their size, because Chinese tables in early days were only about eight inches high and thus to get under it a dog must be very small, of pug or pekingese type.

The 'ha pa' dogs are said to have been in existence in China from at least 700 BC and were originally known as 'Lo Sze'. These little dogs are plainly the basic breed from which the pekingese was developed, though probably with an admixture of Maltese blood. Early Chinese writings also mention 'square' dogs, 'short' dogs, and dogs with short muzzles; but otherwise there is nothing to indicate what they were like. It is also evident that pariah dogs were present in China, probably in their usual role of scavengers around the cities. The Han Dynasty dog wearing a collar and harness depicted in plate 5 shows evident pariah characteristics and the chow chow also appears to have pariah admixture.

China is a land with a chequered history, periods of great prosperity and culture alternating with times of conquest and despair during which the natural talents of the people were unable to assert themselves. During most of its history, this vast land has been ruled by a succession of dynasties. The Emperors exerted their influence through the wealthy Mandarin class, which provided governors of provinces, magistrates and other officials of bureaucratic control. At the beginning of the Christian era, the Chinese peasantry possessed adequate land and lived prosperous and happy lives. But during the centuries to come there occurred the most frightening population expansion ever known: China became the most over-populated land in the world, and the most ill-used. As the population increased, it became more difficult for the poorer Chinese to keep dogs and significant change came over man/dog relationships.

In area China covers some $3\frac{1}{2}$ millions of square miles, being equal to Europe (including European Russia) and one quarter as large again as the United States. In this area live nearly 600,000,000 people, with a density of 155 to the square mile. These figures include Tibet, where the density is only three per square mile; in Shantung the density is 1,300 and in Kiangsu it is 2,000. During forty centuries these thrifty, hardworking people have cleared for farming every available foot of soil and cut timber and brush for lumber, thatch, fuel and every other purpose for which it is suitable. Less than one-tenth of Greater China can be farmed under present conditions and the people have been hard pressed to grow sufficient food for themselves. Every acre is used, every hillside has been terraced where it can yield even a single yard of soil; even human dung is preserved as manure for the fields, thus contributing to the spread of ill-health and disease.

In spite of their high intelligence and their industry, among the common people food consumption is markedly below acceptable minimum standards. Nevertheless, until the last few years the Chinese people clung to a rugged individualism. Agriculture was of the market gardening type, each peasant clinging stubbornly to the family patch of land where his ancestors were buried and the possession of which was rooted in the familial system.

Under such conditions the average Chinese, apart from the wealthy Mandarins, was unable to keep animals. His love for them was diverted to singing birds and people were commonly seen walking the streets carrying a song bird in a cage. Insects were also popular; crickets in particular were valued for their song and competitions were held for singing crickets. Dogs around the cities were ill-fed and quarrelsome, scavenging for their sustenance as do pariahs in other parts of the world. Those who could afford to keep dogs owned the smaller breeds of pug type which require less food.

That the Chinese liked to own dogs when they could is shown by the once familiar Chinese greeting when visiting a friend's house: 'What is the honourable name of your dog?' This wistful question was asked and answered with the wonderful courtesy so characteristic of the Chinese people, although both host and guest knew that the honourable dog was but a china figure and that it might have been many years since the families of either could afford to keep a dog.

The development of relationships between man and dog cannot, therefore, be traced in China, which made few significant contributions to the natural history of the domestic dog. The exception was the achievement in the imperial palaces of breeding dogs of pekingese type to a height of perfection probably never reached with any other breed. No other, not even the saluki, has been consistently developed over so long a period, representing so many generations of canine lives.

The grace, elegance, and luxury of the imperial palaces of Peking and elsewhere were in marked contrast to the general poverty of the Chinese people. The wealth of the emperors was enormous and in these relaxed and cultured surroundings, as much art and love was devoted to the development of the little dogs as to the finest silk picture or embroidery. No student of Chinese art could doubt that the driving force in the development of these small creatures was the Chinese near-pathological love of perfection in all things artistic.

In very early days the Chinese started the cult of miniature 'sleeve' dogs, which they carried in the sleeves of their gowns (probably partly in order to keep themselves warm). Originally, too, large and fierce dogs were kept as watchdogs and later the dog became the guardian of the household and symbol of the family's wellbeing. The ivory, jade or porcelain 'Fo' dogs were also a religious symbol; they were Buddhist lion dogs and became popular with the introduction of Lamaism from Tibet in the eighth century, the lion being one of Gautama Buddha's symbols. These Fo dogs were made in pairs, the female being represented with a puppy beneath her left foot or scrambling up her left foreleg, while the male is shown with his right foreleg resting on a ball of bamboo (plate 6).

The cult of the pekingese began for the odd reason that there were no lions in China; and so the Chinese, with characteristic persistence, developed a lion dog with which to honour the Lord Buddha.

The Chinese emperors were served by hordes of greedy and unscrupulous eunuchs, each vying with the other for power and imperial favour. One way to achieve this was through the breeding of small lion dogs of outstanding merit. In the imperial palaces these dogs had a position of great privilege. They surrounded the emperor in every office of state: when he went to the throne he was preceded by dogs of special colour which barked to announce his entrance; other dogs followed him and carried the train of his robe.

The penalty for smuggling dogs from the palace was the 'death by a thousand cuts'. Nevertheless, many of these dogs were smuggled out of the palaces and found their way into the hands of wealthy Chinese and they were even exported to European countries.

The mysticism which surrounded these dogs started with the emperors of the T'ang Dynasty, which lasted for 300 years from the beginning of the seventh century. These emperors were especially devoted to their dogs and it was during their era that the marvellous feats of breeding these small animals began. Certainly it was then that the lion dog came to its full development and the pekingese breed emerged as an entity of its own.

The T'ang emperors embraced with enthusiasm Tibetan ideas of the lion of Buddha and developed their little dogs to resemble a lion as closely as they could. It was generally believed that the lion of Buddha could make itself as small or as large as it liked and these dogs were supposed to be capable of inflating themselves to the size of a real lion. The idea of the dog as the 'lion of Buddha' appears to have originated in China and to have reverted from there to Tibet, where lion dogs were bred and closely guarded in the temple precincts.

Thus by parallel evolution, forms of pekingese dogs were bred simultaneously both in China and Tibet; China was the suzerain state, Tibet the spiritual centre. The Lamaist religion was full of mysticism and ritual, based on the old animistic rites tainted with ideas of witchcraft and demons. These coloured the beliefs of both the Tibetan and Chinese peoples, giving them the picturesque associations with winds and dragons which so intrigue western peoples. No records of the generations of lion dogs are known, nor how the different forms of shape and size and colour were developed; this knowledge was, however, retained in the memories of the eunuchs and was put to good use in developing the desired properties.

The extreme pug nose of the pekingese was probably developed from breeds which possessed the short muzzled tendency already noted in Far Eastern dogs. It was developed deliberately by the Chinese and later adopted by the Tibetans, where such dogs were called 'hand' dogs. The nose in these unfortunate animals was often deliberately shortened by surgical means; in western countries it is still being shortened by breeding so that the dogs are liable to become undesirably deformed.

The cult of the lion dog was carried secondarily from China to Japan through Korea with the spread of the Buddhist religion. The Japanese carried the lion dog cult to the same extremes as were adopted in China. The dog concerned is, of course, the Japanese spaniel, derived from early breeds of pekingese. The Japanese, like the Chinese, placed lion dogs as guardians over their homes and buildings.

Chinese history, like the breeding of pekingese dogs, shows no continuity. The Han Dynasty was followed by 400 years of unrest with conquest by barbarians and various short-lived dynasties. Then came the T'ang Dynasty (AD 618–907), the Sung Dynasty (AD 960–1280), the Mongol conquest (AD 1280–1360), followed by the stable Ming Dynasty (AD 1372–1643). During the years of unrest the pekingese strain of dogs evidently survived, probably because by this time they were the property of the common people.

The Ming emperors deserted the lion dog cult, but strangely enough they developed a love of cats. During this dynasty, breeding of the lion dogs was left to the eunuchs and the women of the court, with whom they were still popular; the loss of mystic influence at this time may be accounted for by the decline in Buddhism. The Ming Dynasty was overthrown in the seventeenth century by the Manchus, who proved to be the last ruling house. During the reign of the Manchus, the lion dogs again became of great importance, both for their beauty and personality and for religious reasons. It was during this era that were developed the marvellous breed variations first seen by the western world on its penetration to the Chinese capital. The cult of the lion dog as a religious symbol was revived and the lamas again sent gifts of lion dogs to the emperor, thus indicating his identification with the Son of Heaven, the Lord Buddha. These emperors sincerely believed in their kinship with the Buddha himself and the last Dowager Empress was well known to the world as 'Old Buddha'.

Although they had expelled the lamas from China, the Ming emperors could not entirely dissociate themselves from the lion dogs, no doubt worried in case their infidelity might be visited on them. It was in this era that the finest flowering of porcelain art was developed and lion dogs modelled in clay or painted in exquisite silk pictures were substituted for the real thing. Such treasures were stored in the fairyland of the Summer Palace, situated six miles distant from the Forbidden City of Peking. The eyes of the western world first fell on these treasures and on the sacred dogs when the Summer Palace was sacked and looted by the allied troops in 1861. Some of the world's most priceless treasures were destroyed during the looting, in which hordes of Chinese rebels took part. European nations, exasperated by the treachery of the Chinese rulers, entered the capital of Peking. Infuriated by the treacherous seizure and torture to death of their emissaries to the emperor, the allies sacked the Summer Palace and destroyed it. Unfortunately they were unable to keep control or discipline among their own troops or among the Chinese and so was committed one of the greatest acts of vandalism in history.

When this happened, the emperor left the capital with his wives, concubines, and dogs, and took refuge in his fortified hunting lodge at Jehol at the foot of the Mongolian hills. He appears to have taken away almost all of the hundreds of dogs that would normally have been present in the Palace. The only dogs that remained were a small group which kept guard over the body of one of the imperial princesses. Too proud to flee with the emperor, she had taken her own life rather than surrender. These dogs, five diminutive 'sleeve' dogs, were brought to England. One was fawn and white, afterwards named 'Lootie', which was given to Queen Victoria by General (then Captain) Dunne. The little dog was so small that it could curl up and go to sleep in a service cap. Its portrait, painted for the Queen in the following year, shows a perfect miniature lion dog (plate 8).

The emperor of China never returned to Peking. The following year his dead body was brought back in a magnificent procession and he was buried with his illustrious forefathers.

The last fifty years in the history of imperial China were dominated by the great Dowager Empress, whose reign coincided almost exactly with that of Queen Victoria. Her son, the Emperor Tung Chih, died in 1875 at the age of 19 of smallpox, and she arrogated to herself an unseeing despotism which she herself was to deplore, together with its consequences, on her deathbed. She died in 1908 and was laid to rest, a model of her favourite lion dog leading the procession before her bier. Her effigy in a boat suffered ceremonial burning; she was surrounded by her favourite lotus flowers, life-sized figures of her attendants in robes of silk, as well as models of all she loved and might need in another world, including her dogs. Thus the last great figure of imperial China, 'Old Buddha', went to her ancestors surrounded with the panoply of Neolithic burial, leaving the sleeping giant of China to seek her destiny in a modern world.

After nearly 2,000 years of breeding in terms of human generations, perhaps 200,000 years in canine generations, the little lion dogs made their entry to the western world. They have the matchless quality of the finest in Chinese art, expressed not only in their beauty, their proud bearing, their pluck and devotion, but also in their regal character inherited from such a long imperial ancestry. It is these qualities which make these little dogs unique.

During the many centuries of his lineage, the pekingese was bred in various forms, sizes, and shapes. It is not the purpose of this work to describe these variations. Rather is it our object to seek in the natural history of man's companion, *Canis familiaris*, the ways in which he has contributed to human society, as well as the ways in which human influence has contributed to the various ways of life of dogs. The story of the Fo dog of China is unique, not only in the length of time in which these animals have been selectively bred, but in the nature of the association between dog and man. The symbolism of the lion in the dog, based on the marriage of the sophisticated religion of Buddhism with an animistic worship of winds and spirits and devils, with the lion dog in the central place as a symbol of the Lord Buddha and guardian of family and property against spirits, is surely unique. In some respects it is similar to the position of dogs in the Aztec world.

It is of great interest to note the strange resemblance between the little dogs kept by the Aztecs and the 'ha pa' dogs of China. The more primitive inhabitants of the American continent, comprising both Eskimos and Amerindians, domesticated and kept a motley crew of dogs derived from any wild canine species available and suitable for the purpose. In contrast, the two civilized races of the Aztecs and Incas developed little dogs of

53

high quality which were used both for sacrificial rites and for the table. These dogs had special characteristics unlike any possible canine ancestor existing wild in these areas. According to Diaz del Castillo [20] dogs used for food were castrated.

In breeding dogs, the Aztecs were activated by mixed and apparently contradictory motives, which led to the development of the smallest breed of dogs in the world, the chihuahua. The story shows parallels with the development of the small dogs in China and it is tempting to speculate whether similar traditions in China and Mexico were inherited from common Mongolian ancestors. These would have diverged in very early days, the one group spreading southwards into the Chinese plains, the other crossing by the land bridge at the Behring Straits and thence moving southwards until they settled on the lovely plateau of Mexico and the mountains of Peru. Probably, however, this is a case of parallel evolution amongst two peoples having a common ancestry and background.

The Aztecs are said to have kept several breeds of dogs, though early records do not clearly differentiate between them. One authority, the sixteenth century Spanish friar Bernadino de Sahagun, gave seven different names of native dogs. One small variety was specially fattened for eating and was said to be excellent. Dogs were also used as beasts of burden, and dog hair was used as wool. A Mexican, when he died and was on his way to whatever heaven received early Mexicans, supposedly crossed a broad river on the back of a little red dog. Thus it came to pass that little red dogs were killed and buried with the dead, sometimes mummified. Occasionally, as in China, a pottery dog was substituted for the real thing. As in China also, the Aztecs made pottery figures and carvings of their dogs apparently for mystic purposes in connection with the home or for worship in the temples.

Dogs were apparently raised in the temples for sacrifice and may have been used as scapegoats, taking the place of human beings in blood offerings to the gods. It was also believed that small dogs could be used effectively in curing disease by transferring sickness to other persons. Images of dogs were placed on aloe leaves on pathways and the first passer-by removed the sickness. Dogs were also said to have second sight and to be able to predict the future.

The little dogs of Mexico had an early origin. Pre-conquest Indians are known to have kept them, because skeletons of miniature dogs have been found. They were of very small animals, though probably larger than the six pounds standard accepted today for chihuahuas. Pottery figures of miniature dogs from Mexico have also been dated to very early times.

A grim light is thrown on the purposes for which the miniature dogs were bred, judging by the letters of Cortes to the Emperor Charles V [17]. Cortes unfolds a strange story of an extensive zoological garden maintained by the emperors of Mexico (in his day, Montezuma). This was a private zoo kept in the palace, containing hundreds of birds and beasts and having its own open air lakes stocked with fish and other aquatic creatures. The palace was clearly a building of outstanding size; it was entirely devoted to animals and birds except for apartments for two princes and their retinues.

There were ten pools of water containing different species of water birds, all tame and domesticated. There were pools of salt water for the sea birds, pools of fresh water for river birds; the water was changed at fixed times to ensure its purity and was replenished by means of pipes. Each species of bird was supplied with the food natural to it, fish for marine species, insects and worms, maize and seeds for others. It is recorded that fish-eating birds consumed the enormous quantity of 10 arrobas (250 lbs) of fish daily, which were taken from the salt lake. More than 300 attendants looked after the animals in the zoo, and there were special bird doctors to treat them if they were ill. Over the pools ran galleries where the emperor and his suite would walk and watch the birds in their pools. The palace also contained an apartment for dwarfs, giants and albinos. In the gardens of the palace was another building with large halls; those on the ground floor were filled with cages made of heavy timber, in which were kept large numbers of 'lions', 'tigers', wolves, foxes, and a variety of other animals of the cat family. Presumably the animals described as lions and tigers would be jaguars, pumas, and other South American Felidae.

Two points in this story deserve comment. First, people of such experience with animals would have little difficulty in breeding small dogs of high quality of the chihuahua type. Secondly, the zoo contained wolves among the exhibits and in this respect the Canidae played an honoured part; but alas ! the dog himself occupied a humbler role. Among the chief articles of diet given to the carnivorous animals were dogs. Although fed to the snakes and serpents as well as to the numerous flesh-eating animals, however, the dog was in very good company because among the other chief articles of diet were the bodies of human beings.

The Aztec civilization, although in many respects barbaric and cruel, in other ways was amazingly sophisticated and artistically advanced. Small dogs have been widely bred in all places where sophisticated man has settled, though for the most part their development was not aimed at useful or commercial ends. Among the Mexicans alone the uses of these small dogs were non-altruistic, namely for food, burial rites, magic and medicine.

Before improvement and crossing with terrier stock in the United States, the chihuahua must have been a surprising little animal. It is recorded that this breed became feral in the mountains of Mexico and was able to support an independent existence, feeding on young birds, small rodents and other natural food. That so small a dog could do so is indeed surprising.

Part Two

Races of Dogs

Chapter 5

The Dingo Group

THE dingo group of dogs is distributed throughout the Near and Middle East and is widespread in Far Eastern countries. Dogs of similar type are distributed all over Africa. The dingos were introduced to the Australian continent by the aborigines, who went there during the last phase of the last glaciation when the sea level was low. The dates at which dingos were introduced to the Australian continent therefore could not have been much later than 15,000 BC, and this places the date by which some dogs were certainly semi-domesticated.

The Australian aborigines at that time can hardly have emerged from a late Palaeolithic mode of existence; they were nomadic hunters without agriculture, herds of cattle or other domestic animals, and their primitive way of life is little advanced on this today. Once established in the Australian continent, the dingos reverted to a feral existence, living as predators on the abundant marsupial fauna. Following immigration into Australia by white races, they became a serious pest because of their attacks on sheep and they have now been largely eradicated except in small areas around the aboriginal settlements. Although most of the dingos were wild or feral, at any rate until recent times, most aboriginal families have their own dogs which they value and which are well cared for and used in hunting. We have a situation, therefore, which may well have existed in the early days of canine domestication when the same animals existed side by side both in the wild state and in conditions of domestication.

The dingos themselves differ little from the Asian wolves, *Canis lupus pallipes*, which without doubt are their direct ancestors. They are perhaps slightly smaller than these wolves, though their legs are somewhat longer. The average height at the shoulder is 24 inches. The tail shows the primitive wolf-like characters, being bushy and carried hanging down over the anus and between the legs. The ears too are wolf-like, being held erect and forward. When the dingo runs, it runs like a wolf with the head erect.

The long body hairs are generally yellow or white, and there is a greyish underfur. The muzzle is often black and sometimes black varieties of the dingo are found. There is a tendency, as with the pariah and African 'shenzi' dogs, for the end of the tail to be white.

The dingo, in fact, is indistinguishable from the ancestral wolf except by relatively unimportant features which may be within the normal range of characters of the wolves themselves.

Dingos usually hunt by night in parties of five or six; often a mother will go hunting together with her puppies. Occasionally troops of 80 to 100 dingos have been seen. Dingos, like other wild animals, have a strong territorial sense, both defending and respecting each other's territories.

Dingos come into season only once a year, unlike western breeds of dogs which come into season twice a year. The dams usually make nests in holes in hollow trees and litters of six to eight puppies are born. Dingos interbreed freely with European races of dogs, so that pure breeds are not often found today. However, like all members of this group, their chief characteristics are stubbornly retained and they have not divided into numerous breeds like western dogs. This is a strong argument for a single as opposed to a multiple ancestry. In the wild, these dogs do not bark, but they can learn to do so from other dogs. As with the pariahs and shenzis, the hairs of the back tend to be raised in a ridge when the animal is frightened or offended.

Pariah/shenzi-type dogs are widespread in Africa, in the Mediterranean basin, southern Asia, Malaysia, Oceania, India and other eastern countries. They tend to be more stocky than the dingos and show fewer primitive characters. In particular, the tails are turned up on their backs and are short and curled, not bushy. These dogs are nearly always in a savage or semi-domesticated state, and like the dingos have gone secondarily wild in many places. Whereas the dingos have remained in association with human beings still in a hunting/gleaning stage of development, the unfortunate pariahs, in association with more advanced peoples, have become dependent on scavenging the refuse of city dwellers. The extent to which these dogs have taken to a feral existence has suggested to some authorities that the pariahs of Egypt are indeed wild canine species and they bill them for the role of ancestral dog intermediate between wolf or jackal and *Canis familiaris*.

The cringing and scavenging habits of the pariah/shenzi dogs readily suggest jackal characteristics; and since jackal and dog readily interbreed and produce fertile offspring, it is more than probable that some jackal blood may have appeared in them. As with the dingos, shenzi dogs readily become feral and in parts of Africa they are found hunting small game in organized packs. In places they become a nuisance because they kill the young of wild antelope and gazelles.

Pariah-type dogs were well-known in ancient Egypt and, as already mentioned, superior breeds resembling them, not unlike the modern basenji, are to be seen in paintings and sculptures on the royal tombs. There is no evidence that pariah dogs had become a nuisance around ancient Egyptian cities, but evidently they were so in Palestine in Biblical times.

In many parts of Africa shenzi dogs are valued by their owners and are well fed and cared for. This occurs especially where they are required for hunting. Where they are well kept, they are handsome creatures, stocky and powerful, and pleasant tempered. Today, probably the majority of shenzi dogs in Africa have been crossed with western breeds and pure dogs are hard to find. Their African owners are not slow to take a bitch on heat to the vicinity of a foreigner's residence where a good dog is kept, in the hope that their bitch will be served. Thus an improvement in the general standard of the shenzi dogs is evident, although, like the dingo and all members of this race, they stubbornly retain their original appearance and general characteristics.

Around most native settlements, the dogs appear undernourished and diseased, are frequently mangy, and often covered with ticks. Like the eastern pariahs, they are usually snappy, cringing, and resentful.

Breeds derived from the dingo group of dogs are two in number, both coming from Africa. The Rhodesian ridgeback has arisen as a result of crossing native shenzi (kaffir) dogs with imported European breeds over a period of some centuries, ever since the original settlers went to South Africa. From these crosses, this breed has been developed chiefly for hunting purposes. As with most crosses involving this race of dog, the characteristics are predominantly pariah, although the dog is much improved. The other derived breed of this group is the basenji; they are very similar to dogs of the same race which were popular in ancient Egypt. They were rediscovered in recent times in the Congo and Sudan, since when their characteristics have been described for Kennel Club purposes and the breed is accepted for showing in Great Britain. Like many of the dingo group, the basenji are silent.

Above in outline is the history of this interesting group of dogs which comprises about half of the world's population of *Canis familiaris*. To the student of domestic dogs, the group is of great importance, though tantalizing because of the dearth of literature from which facts can be acquired.

Their importance lies in the relationship developed between them and pre-agricultural tribes from northern India to Australia, and the possibility this gives to fix a tentative date by which this association must have come into being. This question is examined by Zeuner [70] in his well-known work on the history of domesticated animals, who states that the introduction of the dingo into Australia is not very recent. Neither its living presence nor its remains are known to occur in either Kangaroo Island or Tasmania. It is therefore moderately certain that its advent was after the separation of these islands from the mainland, suggesting that it was after their occupation by the aborigines. On the other hand, Zeuner quotes the well-known authority Professor F. Wood Jones [30] who puts forward reasons for believing that dingos were brought to Australia by the aborigines when they first arrived in the country. For geological reasons this event must have taken place at least as far back as the last phase of the last glaciation, when the sea level was low, and the dingo is therefore at least as old and probably older than the Mesolithic dogs of Europe and Asia. In addition, dingo bones have been found in the Australian continent together with those of some of the extinct marsupials; it is known that some of these have become extinct in comparatively recent times so, while suggesting an early presence for the dingo in Australia, it does not necessarily imply extreme antiquity.

In the hands of the aborigines, the dingos are at best semi-domesticated, and it is reasonable to regard this type of association between dog and man as being extremely primitive. These people live lives resembling those of Upper Palaeolithic peoples, perhaps when they were emerging from a pure hunting economy to one of hunting and gleaning, and the present day association between aborigine and dingo presumably represents that which existed at the time when wild canine stocks first came to be connected with human activities in the pre-domestication era.

When white settlers first went to Australia, dingos were feral, living in a way similar to that of wolves, hunting in packs and feeding on the numerous marsupial fauna. The aborigines acquired their animals by stealing puppies and rearing them in their households. As house dogs, they are loyal and faithful and generally as trustworthy as fully domesticated breeds of dogs. They were in addition much prized for the services they could render in hunting.

A tantalizing question is whether the original dingos taken to Australia by the aborigines were domesticated animals; that is to say whether they had been improved at any time by selective breeding, or whether they were unselected stocks derived either from the pale-footed Asian wolf, or from some canine stock closely related to it. It is improbable that this question can ever be answered. It may be thought unlikely, however, that primitive man could have stolen the puppies of Asian wolves and used them as freely as they do the dingos; if they did so, the resulting stock taken to Australia would be expected to have more the conformation and characteristics of the wolf and lack those which are so peculiarly characteristic of the dingo. The alternative, and possibly the true explanation, may be that some members of the wolf stock had of themselves taken to a form of existence in which they relied to a great extent on scavenging around human settlements, just as the related pariah dogs do today. In this way without deliberate breeding a somewhat specialized variety of the Asian wolf could have come into existence by natural selection, smaller in size and with qualities of docility such as would render the association acceptable to man.

Dingo characteristics are so marked, with the triangular forward-looking face, the pricked ears mounted well forward on the head, and the flat brow, that they are easily recognized among the many breeds which exist over the large area they inhabit. The only notable variation within the group lies in the conformation and carriage of the tail, about which more will be said below. We recognize these special characteristics in the little dogs from New Guinea, picturesquely known as the 'New Guinea singing dogs' because of the melodious howls they produce at sundown.

As already mentioned, Zeuner gives a picture of a young dog of typical dingo type (apart from the tail) of the pre-agricultural Chenchu tribe as an example of the pariah dogs of India. He also remarks on the resemblance to dingos which he saw among the pariahs in the remoter parts of northern Gujarat in north-west India and states that these dogs are in almost every respect like dingos – fawn-coloured, sleek-haired and medium-sized, with upright ears. They breed remarkably true to type, deviating only in colour. Pariah dogs closely resembling dingos are distributed from the Balkan Peninsula, Asia Minor and North Africa to India, Java and Japan. Zeuner points out that it is not difficult to understand how early man might have taken such dogs to Australia, where they established themselves rapidly, pariah fashion; they became partly independent of man, since the Australian environment offered them plenty of food with numerous species of marsupials unadapted to the presence of a true carnivore.

The dingo-pariah group of dogs, like the pre-agricultural peoples still living in the world, provide us with an example of living history, taking us back far beyond that which can be achieved by the study of other groups of dogs. In the absence of positive evidence, conclusions must necessarily be tentative. Nevertheless, the antiquity of the breed is fully demonstrated by primitive characters, both in connection with the breeding and barking habits, the absence of dew claws on the hind limbs, and the stubborn uniformity of conformation which has lasted over many centuries. These are well shown by the study of the teeth, as illustrated in plates 36–40; unlike in artificially bred domestic breeds of dog, the incisor teeth are powerful and well spaced as in wolves, to which the resemblance is striking.

As already noted, the tails of dingos are carried low, covering the anus, and are somewhat bushy after the fashion of the wolf. It is tempting to suppose that the characteristic tail of the basenji was deliberately produced by the ancient Egyptians by selective breeding. They appear to have appreciated this type of tail in their dogs. Basenji-like dogs with tightly curled tails are shown on tomb paintings and curly tails are also depicted in greyhounds from early dynastic times, a feature which is quite alien to greyhound stocks today.

Among the many studies that have been made on genetics and inheritance in domestic dogs, very little has been discovered about the inheritance of the different tail features, although this is plainly a factor of great importance. How does it come about that the pariah dogs throughout their great area of distribution from the Balkans to India differ from the parent dingo stock in the possession of these characteristic tail features? One supposes that in wild Canidae the low carriage of the tail serves a purpose in masking the scents from the peri-anal glands when trouble looms; and conversely it enables the scents to be unmasked by raising the tail at times when recognition of one animal by another is advantageous, as when the male seeks a female. The curly tail can only argue that at some stage the pariah dogs have been selectively bred by man throughout their entire area of distribution, as apparently was done in early dynastic Egypt.

The existence of feral breeds of pariahs with dingo-like tails both in Egypt and in the Middle East has recently been brought to our notice by the kindness of two persons now breeding these dogs in the United Kingdom. Mrs Connie Higgins owns a bitch of this breed; and Mr Mold has for some years now been breeding such dogs and has given us accounts of their conformation and characteristics. We have also been able to study photographs of them and Mrs Higgins kindly permitted us to examine her bitch. It is our belief, derived from our studies, that these are pariah dogs which have become feral and, as occurs with other animals to which this happens (notably pigs), they have secondarily reverted to the more primitive characters of the parent stock. Unlike the scavenging pariahs which abound in the Middle East, these dogs combine to hunt living prey in isolated localities of mountain and desert and have thus become, like the dingo, a secondarily wild animal. Like the dingo also, when brought into captivity they are trustworthy and loyal members of the household, with a great attachment to one master. Their resemblance to the dingo is marked, although they are somewhat lighter in build. Like the dingo they run with head held high. They appear to have only one heat period in the year, but some of the pups are born with dew claws on the hind limbs, arguing an ancestry from a dog that has at one time been in domestication.

The tail conformation appears to us to have added significance, enabling the migrations of dingos of pariah type to be traced. We shall establish in the next chapter that one group of the northern spitz dogs, the Finnish spitz, was derived from dogs of dingo type taken by the Finno-Ugric groups of peoples in their north-eastwards migrations to their present lands. It is significant that not only in this group but also in the elkhound group of northern dogs, curly tails resembling those of the basenji and pariah are perpetuated. Tails of this kind are not characteristic of the northern dogs as represented by the huskies and shepherd dogs of the collie and alsatian types. They are, however, characteristic of the elk-hounds, which in other respects very much resemble dogs of alsatian type; on the balance of evidence, they appear to be derived from the northern wolves rather than from the Asiatic wolves through dogs of dingo type. However, some dingo (pariah) admixture may well be indicated through the Finnish spitz, judging by the tail conformation.

The known presence of dogs with typical dingo characteristics in northern areas lends strong support to the probability of such admixture. Are such affinities possible in other breeds which show similar tail characteristics? They are present also in the Maltese dogs of today, but were not depicted in ancient Egypt and Greece. It is unlikely, therefore, that dogs of peking-ese type acquired such characteristics from early admixture with the Maltese. Nevertheless, the tail curl of the pekingese does suggest the possibility of pariah blood at some stage. The existence of true pariahs in Japan suggests that they were also present at one time in China, and evidence of this is surely seen in the chow chow, of which the facial and tail conformations are both suggestive of partial pariah origins. This dog is usually supposed to be derived from samoyed-type dogs, of which the colouration and tail characters also suggest the presence of pariah blood.

The study of the dingo-pariah dogs is in one sense made easy because they are distributed in more or less pure form over a vast range from North Africa and the Balkans to Japan, the Indonesian archipelago, and

Australasia. It cannot be doubted that from very early times, admixture of this basic breed with others developed in different areas has had a profound influence on the conformation of many of our modern breeds. The characters on which one may attempt to trace the distribution of mixed dingo or pariah blood are so tenuous that they can plainly have no final validity. Possibly a study of canine blood groups could enable some more precise estimates to be made of the extent to which the different parent stocks have contributed to the breeds today. Such studies have enabled estimates to be made of the admixture of different stocks in modern human races. Precise experimental work on similar lines would certainly be rewarding in relation to *Canis familiaris*.

Evidently in early times a mutually satisfactory relationship had developed between man and this primitive dog to the advantage of each; as a competitive predator, in Australia at any rate, it was not resented and having been taken into captivity in puppyhood, it became an accepted member of the tribe. In early Egyptian times, these dogs were household pets and used for hunting; they were valued, and selectively bred. Such a situation seems to have continued at least to the time of the Ptolemies, as is evidenced by the skull depicted in plate 40 of a pariah from the canine cemetery at the Temple of Denderah, which was built by Cleopatra.

In later times, the relationship between man and dogs of this type presents a sad story. It is generally supposed that the true dingos are untrustworthy and do not make suitable household pets and there is a hesitation, probably unjustified, to offer or accept puppies of this breed. Today, however, there is a tendency for the qualities of dogs of dingo type to be appreciated again. Both the basenji and the Rhodesian ridgeback have found favour; and Mrs Higgins' and Mr Mold's wild pariah breed could well achieve popularity if they are able to establish and distribute them.

Chapter 6

The Northern Group

IN spite of the lack of literature regarding conformation and local varieties, the canine historian has in the dingo/pariah group concrete evidence on which he can base reasonably sure assumptions. He has a numerous and homogeneous race of animals, spread over a large portion of the earth's surface, and he has certain yardsticks by which he can apply some sort of chronology. He also has in *Canis lupus pallipes* a good candidate for the role of ancestral wolf. Finally, these dogs have certain rather striking characteristics, from which it is possible to suggest their probable routes of dispersal from the main geographical areas and the way in which they have mixed with other breeds. The centre of domestication for the dingo is likely to have been that area where tribes of Australasian stock existed, namely throughout the Indian sub-continent. The greyhounds were probably developed in southern desert regions. The mastiffs appear to have been developed in the mountainous areas which run from the Himalayas to the Pyrenees. The fourth group, that of the northern dog (of which a number of breeds are known as 'spitz') we believe to have been developed primarily from the large grey wolves of northern Europe, to which some breeds bear a close resemblance. It is, however, abundantly evident that in most breeds there has been much mixing of these dogs with those of other groups. There are at least two types which we place in this group of which the true ancestry must be somewhat dubious: one consists of shaggy sheepdogs like the Russian owtchar and the old English sheepdog; the second comprises the important and popular group of the terriers.

We have already referred to the northern movement of a group of pariah-type dogs with the Finno-Ugrian peoples; these became ancestral to the Finnish spitz. We have suggested that pariah characteristics can be traced in such dogs as the elkhounds, the samoyeds, and the chow chow, all of which are classed among the northern spitz breeds. Other northern breeds such as the Karelian spitz and laika, and the Alaskan malamute, also show characteristics likely to be derived from dogs of this type. According to Kathleen Risbeth, MA[1] until lately librarian at the Haddon Library of Archaeology and Ethnology at Cambridge, during Neolithic times the Ugrian tribes were living in the Volga region of Russia; from there they began moving steadily northwards until they reached and colonized Finland and adjoining parts of Russia, other Ugrian peoples becoming the Lapps and Samoyeds. Evidently these peoples possessed dogs of pariah type and during Neolithic times they introduced them into northern regions.

The belief has been expressed by Zeuner and other authors that all dogs are derived from those of dingo type or alternatively from some form of *Canis lupus pallipes*. The Mesolithic dogs of Denmark, the so-called Maglemose dogs, are found in connection with human settlements and were undoubtedly fully domesticated. Even at that remote period they were of two different types, one large and one small. The largest were still of smaller size than the local northern wolves, and from this fact it is argued that they must be descended from the smaller wolf breeds known to be ancestral to the dingo group. It is suggested that the characters of the northern wolf were bred into these dogs by crossing bitches with northern wolf males. It is often stated also that the savage, untrustworthy nature of the northern wolves makes them unsuitable candidates to be the ancestors of domesticated animals, which would come in close contact with the women and children of the settlement. Zeuner himself seems little impressed with these arguments, and in another section of his work puts forward the contrary view that the northern dogs were developed from the local wolf populations.

The Maglemosean deposits, on which these arguments are based, date from about 8000–6000 BC. While there is no proof, there is strong probability that the origins of domestication in northern areas must be sought long before this time. The evidence, such as it is, lies in the presence of canine coprolites, that is fossilized faeces of dogs, in connection with late Palaeolithic settlements; and in the presence of animal bones showing tooth marks supposed to have been made by dogs in the settlement. Against this it is argued that the coprolites may be those of wolves which visited the settlement after man had departed, and that the wolves may have gnawed at the bones.

In the previous chapter we suggested that, prior to domestication, the Asian wolves from which the dingos were derived may have developed

[1] vide *New Chambers Encyclopaedia*.

some association with nomadic human tribes, and that these animals may have had the properties of small size and docility which would not be resented by their human associates. Such an association developed over long periods could well produce a strain of the wild wolf stock which differed in important respects from the ancestral animals from which they were derived.

As mentioned in chapter 2, both the northern and Asian wolves will, if necessary, obtain their livelihood by scavenging rather than hunting. When man was a nomadic hunter in late Palaeolithic times, it is inconceivable that the wolves would not have fed on what was left after a big kill; indeed the dingos in Australia do this, and they also clear the refuse in the aborigines' settlements. Having learnt that there were good pickings to be gained, wolf packs would no doubt follow in the wake of bands of hunters. Furthermore, when man began to live in more permanent settlements in Mesolithic times, no doubt wolves would enter at night to clear the refuse and would not be resented for performing this service. It is thus abundantly evident that with northern wolves domestication could have been a two-stage affair, as it probably was with the Asian wolves.

The arguments against the direct participation of northern wolves in the ancestry of the northern races of dogs therefore cannot hold good; and indeed the conformation of these dogs, except in cases where pariah participation is known to have taken place, supports this conclusion. There are no reasons for refusing acceptance of the northern wolf as the primary ancestor of these dogs. Statements that these wolves are not sufficiently tractable to be domesticated are known to be erroneous; on a number of occasions they have been domesticated and when raised from cubs they are both safe and loyal. Mr and Mrs Crisler, in a famous and much quoted case, raised a young wolf cub in Alaska and had as good a pet as any domestic dog. Furthermore, as already mentioned, the Amerindians domesticated many species of wolves and wild dogs present in North America, almost certainly including the Canadian timber wolf, which is virtually identical with the northern grey wolf of Europe.

It is just possible, though probably untrue, that the foundation stock of northern dog breeds was derived from some early northward migration of peoples with dogs of dingo type and that these were so extensively crossed with the northern wolves that the dingo characteristics disappeared. In this case the dingo blood has been so diluted that it makes no contribution to the present breeds of northern dogs.

By Neolithic times, around 3500 to 2000 BC, northern dogs had become well established into at least two breeds which were quite distinct. The larger of these was typical of the northern sheepdogs of alsatian (German sheepdog) and collie type, which by Bronze Age times were very numerous. Small breeds were developed in the Neolithic lake dwellings of Switzerland, Hungary, and parts of Russia; probably the reason was that, as in city flats, large dogs would be ungainly companions in a small space. The skulls of some of these dogs were no more than 125 mm in length, that is under four inches, so that they were very small dogs indeed. At this time, too, it appears that dogs of spitz type were finding their way by Neolithic trade routes to the Mediterranean world, where the Maltese dog of spitz type was to become a firm favourite for many centuries.

In general, the northern dogs segregated into two definite types, the long-muzzled and the short-muzzled; these types are well seen in the huskies with short muzzles and a pronounced 'stop', and the long-nosed collies with virtually no 'stop'. There is a similar difference in conformation in the short-muzzled mastiffs and the long-muzzled greyhounds, although in this instance it would appear that the breeds were derived from different ancestral groups of wolf. That this tendency is already present in northern wolves is shown by the striking pictures of muzzle conformation in two wolves from Whipsnade Park (plate 36). The implications of this are discussed in a later chapter.

Zeuner makes the important point that primitive man had no need to train his sheep dogs to round up herds of animals and to separate the ones he needed, because in their hunting the northern wolves already did this on their own account. He reproduces an outstanding aerial photograph which shows a single wolf separating one animal which had been selected from a herd of caribou; this is reproduced by courtesy of the Canadian Wildlife Service in plate 16.

With the northern wolves, we have no milestones of time which can help us to determine the dates when man domesticated them in the sense of taking them into permanent captivity and selectively breeding them for his purposes. The uniform climate of the last glaciation had existed for tens of thousands of years and the lives of Upper Palaeolithic peoples must have been uniform, lacking external stimuli to make changes. Man's numbers were small, the prey on which he and the wolves lived abundant. Man was nomadic and presumably he lacked the inclination and the physical ability to breed and rear the wolves as domestic animals. It is most plausible to believe that an association similar to that of the Australian aborigine with his dingo stocks may have developed, namely that cubs would be captured and taken into partnership with the tribe. When grown, they might perhaps be used in hunting and in drawing sledges of skin. As with the Samoyeds today, the flesh would probably be eaten and the skins used for clothing. These people appear to have used a form of ski to travel over the snow; they also had sledges made of leather on which they would transport their family goods, or carry the animals they killed in hunting.

There was probably a long tradition, which could have extended over many thousands of years, of using wild wolves in a semi-domesticated way. True domestication was certainly developed during Mesolithic times and was the result of the changes which affected the lives of northern human beings at this stage of development.

In response to new needs, the reindeer, which had been hunted for so many years, was taken at least into semi-domestication. At the same time also, dogs evidently began to have more varied uses, though they would still have been used for drawing sledges and for hunting. For instance, their services in clearing the food debris from the settlements of peoples who were no longer nomads would of itself lead to a greater intimacy and to their acceptance in the home, where they would bear their litters and thus lay the foundations of selective breeding.

This combination of circumstances led to the development of different breeds of dogs, of which there were three main types: the sledge dogs; the hunting spitz dogs, such as the elkhounds; and sheep dogs of alsatian and collie types. It appears that sheep dogs and shepherd dogs were independently developed also among the mastiff group, and this ancestry is therefore confused. A further group of small dogs was developed which was ancestral to the Maltese dogs, and probably also to our modern terriers.

From time immemorial northern people have relied for transport on sledge dogs, directly derived from northern wolves. Travellers in polar regions argue hotly about the merits of the different breeds, some of which contain a motley group of animals and can hardly be regarded as breeds at all. They are still close to the ancestral stock and have the reputation of being somewhat unreliable as working dogs and untrustworthy in temperament. They retain characteristics of the wild animals from which they were derived as regards both territorial instincts and social hierarchy. Plates 17 and 18 show a team of husky dogs pulling a sledge in the traditional Eskimo way; the second picture shows the outcast of the hierarchy, who was not permitted into the team by the other dogs and who slunk along behind at a distance of some yards.

Native dogs of this type are the Siberian huskies and the Greenlandic Eskimo dogs from Baffin Island. For these duties, however, many northern peoples have deliberately bred more disciplined animals in which dingo admixture usually seems evident. Such breeds are the Alaskan malamutes, the national dogs of Alaska; the Russian laikas; the nootka dogs of Iceland; and the samoyeds.

In spite of their reputation for indiscipline, many explorers would not use dogs other than huskies in arctic conditions. These dogs have fantastic hardihood. They must be able to sleep out in polar conditions and to travel without food for days on end, during which time they pull heavy loads and lose a great deal of weight. Often on return to base their condition is so poor that difficulty is experienced in restoring their condition and they may die of diseases such as pneumonia. The normal food given to them is seal meat, which must be fed to them in large quantities during periods of rest so that they have adequate reserves for the ordeals ahead.

With all their virtues as sledge dogs and sheep dogs, the northern dogs have not produced any breeds which are pre-eminent as hunting dogs, except when mixed with dingo blood as in the elkhounds. The large, primitive elkhounds still exist in various recognized breeds. The Lapland spitz is quoted as being possibly similar to the ancestral type, but the merest glance at this animal shows it to be typically a northern spitz with dingo cross. It is said that in earlier days it was of purer blood, but today it is crossed extensively with other breeds and its abilities have deteriorated. This may well be so, since it appears that its original function was herding rather than hunting, and the crossed elkhound types do not seem to have any particular qualities as herding animals.

The elkhounds proper consist of a number of varieties, including the black Swedish elkhound and the Karelian spitz, which is very similar to the Russian laika but is used exclusively as an elkhound. There are a number of other spitz breeds, such as the Norwegian spitz or Lunnenhund which is a miniature elkhound used for hunting puffins. The Norwegians have another miniature spitz breed called the Buhund, which was developed as a watch dog and later used as a sheep dog. Small dogs are especially suitable in Norwegian conditions. The Västgöta spitz of Sweden is very similar to the Welsh corgi, to which it is believed to be ancestral. It is said that the Vikings brought some of their small spitzes to Wales, where the corgi-type dog became popular as a house dog and guardian dog in the villages. The corgi is a typical spitz dog of the smaller elkhound type; another such is the Dutch breed of spitz known as the keeshond, which has been used for centuries as a watch dog in Dutch villages and, like the schipperke, served as a ship's dog on the barges.

In this group must also be included the chow chow or Chinese spitz, undoubtedly derived from the samoyed. This is a very ancient dog and, like many things in China, has lost its original use and become endowed with mystical significance. It was used in early times in temples to ward off evil spirits, its terrifying appearance being adapted to this purpose. The Chinese also fattened it for food and animals used for butchering were fed solely on a vegetarian diet. The chow chow is peculiar in that there is a blue pigment in the tongue and lips, a character which has arisen from a rare mutation.

The small dogs of the lake settlements of Switzerland and other countries, which have been given the variety name of *Canis familiaris palustris* (turbary dogs), were similar to the smaller spitz elkhounds. Pomeranians and Maltese dogs seem to have been developed from dogs of this type. The Maltese breed can be traced back to early Egyptian times. A model dog of this type from the Fayum, now in the British Museum, is dated to 600–300 BC. They are mentioned by Aristotle and Timon of Athens and are illustrated on a Greek vase of 500 BC. Pliny also describes Maltese dogs in AD 23–79. Apparently they were first imported into England during the reign of Henry VIII and it was a dog of this type which followed Mary Stuart to the scaffold. Today they are especially popular in Canada, and in spite of their Mediterranean derivation they do very well in cold climates. This property may perhaps be attributed to their northern spitz origin.

The pomeranians have diverged from the Maltese and appear to be closely related to northern dogs of elkhound type such as the keeshonds, the modern breeds being derived from German spitzes. In Graeco-Roman times, small dogs of early Maltese type were traded along the spice route with China, and it is very likely that their blood has contributed to that of small Chinese dogs of pekingese type.

It has been suggested, with a high degree of probability, that the turbary dogs of the European lake settlements are ancestral to the schipperkes, the barge dogs of Holland. Some dog allied to the schipperke has also been suggested as the most probable ancestor of the terriers. Their true origins are somewhat mysterious, but a derivation from the smaller elkhound spitzes seems to be most probable.

Even if undisciplined as sledge dogs and undistinguished as hunters (except in hybrid form), as sheep dogs the northern breeds are pre-eminent. There are two main groups: unspecialized dogs such as the alsatian (German sheep dog); and the highly bred long-nosed sheep dogs of collie type. Of the unspecialized types, the alsatian is the best known; but similar dogs are distributed widely and there are many breeds, varying slightly in size and colour. It is often thought that these dogs were only recently bred from wolves, a belief which is far from the truth. They have existed in their present form certainly for 2,000 years and, in spite of their superficial resemblances to the northern wolf, in point of time they are far removed from them. It is said that up to some 30 years ago German shepherds would tether bitches to get them covered by wild wolves, but this is probably untrue. Of great intelligence and devotion, they have found favour particularly for police work in many countries.

The long-nosed collies are closely related to sheep dogs of alsatian type and have clearly been derived from them by selective breeding. For skill and intelligence in handling animals they are unsurpassed and there is no finer sight than these dogs at work rounding up sheep or cattle, obeying their master's distant commands, and using outstanding initiative in under-taking difficult manoeuvres without unduly scaring or damaging their charges. The collies are an ancient breed of sheep dog selected for these special qualities and for their intelligence – qualities which, as we have seen, were already present in the northern wolves.

Among the northern peoples, the prime requirement in a sheep dog was to drive and control animals, whether reindeer, cattle or sheep. In other areas, however, there was an added need to protect flocks or herds from wolves and other predators and from bandits. A distinction is thus made between *sheep-* or *herd* dogs such as the collie, and *shepherd* dogs, whose main function would be for herd protection. The collies are pre-eminently sheep dogs, but would be of less value in protecting flocks from marauders. It is probably for this reason that in many areas dogs of alsatian type have been retained as sheep dogs rather than the more efficient collies. In more southerly areas, sheep seem to have developed the habit of following the herdsman and the main accent has been on flock protection. Thus through-out the mountainous areas of Europe and Asia, the dogs used for controlling flocks and herds are of a different type – large, fierce creatures, derived not from the north, but from the very large and fierce mastiff stock.

Confusion often arises between 'sheep' and 'shepherd' dogs, which have quite dissimilar origins. Shepherd dogs, such as the Pyrenean mountain dogs, are valuable for protecting the animals committed to their charge, but have few talents when it comes to rounding them up and controlling them. There was thus a need for some form of sheep dog which could perform the dual function of protecting animals as well as controlling them. Such breeds exist in large, shaggy sheepdogs such as the old English sheep-dog, the Russian owtchar, and the two enormous Hungarian dogs, the komondor and the kuvasz. The kuvasz is said to be the largest dog in the world; from its appearance it is predominantly of the *shepherd* dog type, not unlike the Pyrenean mountain dog and others of its kind. The komondor, which resembles it, is nearly its equal in size but has many of the characters of the shaggy sheep dog.

The origin of these dogs is obscure, but presumably they were developed in ancient times by crossing shepherd dogs of mastiff origins and dogs of northern sheepdog type. This has resulted in breeds of sheepdogs which are slower workers than the collies, but still efficient in rounding up and controlling animals, and which also have the size and fierce temperament to protect the herds. Their shaggy coats give them warmth in the cold winters of the Russian steppes, where they were originally bred, and the long hair over the eyes protects their sight from snow.

It is a characteristic of the northern dogs that they become devoted to one master and, in their original form, were untrustworthy with strangers. This is true of the alsatians, collies, and old English sheepdogs. These undesirable characteristics have been largely eliminated in the modern show breeds which are now popular as house dogs. Nevertheless, in contrast with breeds of mastiff origin, they still incline to show individual attachments; this has been discussed by Konrad Lorenz [40].

The terriers are most probably derived from spitz dogs of elkhound type and may be descended from the small dogs of the Neolithic pile dwellers through some such form as the Dutch schipperke. These small, sporting dogs would follow foxes or badgers into their holes and had an especial use for hunting small vermin such as rats or mice. From these beginnings, numerous breeds have been developed for a variety of uses. They have been crossed with other breeds of different origins, to establish yet other breeds for an even greater variety of purposes.

Most breeds of terriers are more or less confined to Europe, though there is the anomalous though not unusual position that they are also present in Tibet. They thus show the same discontinuous distribution that will be noted with dogs of spaniel and other types. This may be due to separate evolution in two different areas, or to communication between Tibet and the western world in ancient times.

Today there are a great many varieties of terriers in the western world; strangely enough, most of them are Scottish, English, or Irish. Some of these breeds are ancient, others have been produced by deliberate selection and crossing in recent times. Many terrier breeds have been named after the localities in which they were developed and it would seem that, as with sheepdogs, they evolved in relative isolation, thus deviating from the original norm. Many, such as the sealyham (developed in comparatively recent times), have been named after the place or state where they were originally bred.

Although the terriers conform to a general type, there are many variations in the kind of hair – wirehaired or smooth-haired; in colour; and particularly in the size and length of leg. The original terrier was probably wirehaired and of the black and tan variety. Many terriers have been crossed with other breeds, and it is suggested that the fox terrier was derived from crossing terrier with foxhound. The bull terrier is the result of crossing the fox terrier with the English bulldog, producing an animal tenacious of purpose but quicker in movement than the bulldog, with certain qualities desired for bullfighting and bullbaiting.

At the time of the Roman conquest of Britain, there was considerable export of British dogs to Gaul and to Mediterranean countries. They were much prized because of their fine scent, although they were said to be rather rough in appearance. It is sometimes supposed that these dogs were the English mastiff, but their description would hardly apply to a dog of this type. In addition, so many excellent dogs of mastiff type were available in the Roman world in the Hyrcanian, Molossian and Germanic mastiffs that it is difficult to see why another should have been so prized. It is more likely, since we know there was continued breeding of good terrier types in these islands, that the dog in question was the old English terrier from which our modern breeds have been derived.

Possibly the terrier existing today which still retains most of the original characters is the border terrier. These dogs have been bred more or less unchanged for countless generations in the Cheviot Hills on the border between England and Scotland. It is still a hunting terrier, large enough to be able to follow a mounted huntsman and yet small enough to get into any fox or badger earth. One of the local Cheviot varieties is the dandie dinmont; one of these is shown in a picture by Gainsborough of 1770.

The schipperke of Flanders is of terrier type; it has fulfilled the function of watchdog on canal barges for a very long time and the name means 'little skipper'. It is probably similar to *Canis familiaris palustris* of the Neolithic lake settlements and may be the link between the true spitzes and the terriers.

There are also Welsh, English, and Irish terriers from which a great variety of forms have been developed by selection and interbreeding. The English fox terriers, both wirehaired and smooth-haired, are probably Durham and Yorkshire dogs; they are of ancient lineage and have been known at least since 1600, although the white colour did not become predominant until the nineteenth century. This was produced by deliberate selection to enable the huntsmen to spot the dogs easily in the undergrowth when they emerged from an earth. These dogs would always accompany the hunt, running with the foxhounds, and would be sent into the earth after the fox if it escaped. To face a fox or badger in its earth would require considerable courage and all terriers are very plucky creatures.

As with most breeds of dogs, the terriers have their toy varieties. The English toy terrier should not exceed eight pounds in weight and another very popular miniature is the Yorkshire terrier. Among the continental terriers with toy varieties are the schnauzers and pinschers. There is a miniature schnauzer and a 'monkey' terrier (Affenpinscher) which is a tiny ratter of sturdy build; it is very courageous and should measure ten inches at the shoulder and weigh seven to eight pounds. Both are German breeds of ancient ancestry.

Chapter 7

The Greyhound Group

THE origin of the greyhounds is lost in antiquity, and there is no clue which could settle decisively the manner of their derivation. They are quite distinct from any of the other groups of *Canis familiaris*. Their special characteristics consist of the lean yet powerful body, the deep chest and the long legs, which give them speed and stamina. In this they surpass all other breeds, just as the bloodhound and its relatives of the mastiff group surpass all others in their ability to find prey by scent. Greyhounds are certainly not to be classed, as is usually done, with the true hounds which hunt by scent, since they pursue their prey largely by sight.

Unlike all other breeds, they are clearly cursorial animals, and their adaptations to this form of life are so marked as to suggest that they are derived from ancestors with cursorial properties. A cursorial animal is one adapted to finding and hunting prey in open country, where it cannot be stalked by stealth but must be overtaken by superior speed and endurance.

All members of the greyhound family have been used for coursing, that is for running down animals, and the distribution of the group has always been in places where open country exists and where this form of hunting was possible. Both today and in ancient times they were found in Scotland and Ireland; but their main area extends by way of North Africa, through Arabia to Russia and Afghanistan; they are not found naturally in wooded areas.

Dogs of greyhound type are depicted in hunting scenes on friezes and paintings from ancient Egypt and Assyria 4000 to 3000 BC. The earlier types were much as they are today, but their tails tended to be feathery or bushy. This suggests that the ancestral form may have had a more hairy and wolf-like tail; the general conformation of the tail also suggests a wolf-like ancestry, for it is carried low though curling upwards, thus covering the anus as is characteristic of wolves.

The ancestor usually suggested for the greyhounds is *Canis lupus pallipes* or one of its varieties. It is quite possible that *C. l. pallipes* at one time produced a cursorial form which could have sired the greyhounds, and Pocock's *C. l. arabs* or a variety of it is a likely candidate. This is suggested by the distribution, which overlaps that of *C. l. pallipes* and *C. l. arabs*. Moreover, in the areas where greyhounds are most likely to have originated, no other wolf stocks are known to have existed.

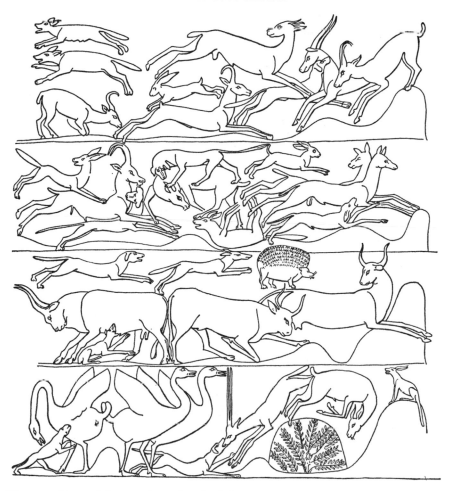

Figure 9. Dogs of the greyhound and saluki type attacking game. Thebes, *c.* 1450 BC.

It is also possible that cursorial wolves allied to *C. l. arabs* were at one time present in North Africa as well as Arabia. Almost within historic times, large parts of the Sahara were comparatively well-watered, carrying herds of wild grazing animals; they would thus be a hunter's paradise for a cursorial beast of prey. Arabia also appears to have had more lush vegetation and probably greater numbers of prey animals in the past.

Pride of ancestry in this group can possibly be claimed by the saluki breed, said by some to be the oldest pure-bred dog in the world, though the pekingese might dispute this. Its origin extends back beyond history and it has been pedigree-bred from time immemorial in the deserts of Arabia, Syria, the Sahara, Egypt and beyond Persia.

Dogs of this type were depicted on tombs of ancient Egypt, such as that of Rekhma Ra, *c.* 1400 BC. The Arabs, with their great love of perfection in animals, bred the saluki with the same care that produced the Arab horses and camels. Their term 'slughi' or 'saluki' means a dog or hound which is valued and respected, as opposed to the 'kelb', the pariah dog of Islam. It has been suggested that the word 'saluki' is derived from Seleucia, a post-Alexandrian Greek empire based on Syria and extending to the

Figure 10. Saluki type hounds from the tomb of Rekhma Ra, *c.* 1400 BC.

borders of India. This suggestion is of interest in view of the age-old belief in Britain that the term 'greyhound' is a corruption of 'Greek hound', although an alternative derivation makes it 'gazehound', a reference to sight hunting. The names associated with greyhounds could well be of Greek origin, although the dogs themselves antedated Greek expansion by many centuries.

The Bedouin, who also have a great love of these dogs, hunt gazelles using salukis in combination with hawks. When the hunted animal enters thick country, the hawk is sent after it and the dogs follow the hawks until they are led to the prey. The dogs are trained not to kill the animal, but to hold it until the hunters arrive and can kill it in the manner prescribed by Mohammedan law.

The salukis were imported into Europe from Persia, but it is known that the Persian breed came from an original home in Arabia. True greyhounds, as opposed to hounds of saluki type, have been present in England at least from the ninth century.

The early greyhounds as depicted on Egyptian tombs are plainly very closely related to the salukis, though they would have been used for smaller game since they are too small to tackle the larger gazelles. The traditional prey which they have coursed since the classical days of Greece and Rome was the hare, for which an electric imitation has been substituted in many places today.

Another member of the greyhound family is the Afghan hound. It was reintroduced to Europe from Afghanistan, but came originally, so it was said, from the Sinai peninsula. This is a distinctive type of the eastern group of the greyhound family. The adults have thick fleecy coats, which no doubt are very necessary in their cold adopted habitat. The puppies, however, do not have fleecy coats and so reveal their origin in warmer climates. These dogs tend to segregate when developed in isolation and vary in size and conformation according to the district where they are kept. All, however, are much heavier dogs than the salukis and are used for hunting relatively large animals. For this purpose they are used in couples, one being trained to attack the hindquarters of the animal and the other the throat, the tactics of the ancestral wolves.

Another eastern greyhound is the Russian borzoi; this breed was evolved east of the Urals, though presumably the ancestors also came from southern Mediterranean areas. In this breed the bitches are considerably smaller than the dogs, a somewhat unusual characteristic in *Canis familiaris*.

These dogs were used for hundred of years by the Russian aristocracy for coursing hares and rabbits and for hunting wolves. Indeed before the eighteenth century they were pure wolfhounds. A description of the methods by which they hunted wolves is as follows:

'The perfect wolfhound must run up to the wolf, collar him by the neck, just under the ear, and when the two animals roll over, the hound must never lose his hold, or the wolf would turn round and snap him through the leg. Three of these hounds hold the best wolf powerless. The men can dismount from their horses and muzzle a wolf and take him alive'.

The suggestion that these dogs were used to capture wolves is of great interest in view of the continually recurring evidence that domestication of dogs was not the result of a single act, but that wolves have been continually caught and tamed, or used for crossing with dogs, over many thousands of years.

Another sight-hunting greyhound of venerable ancestry is the Irish wolfhound. According to the Irish, this hound has existed in Ireland from time immemorial and the facts so far as they can be ascertained would appear to support their case. The Irish wolfhounds, together with the Scottish deerhounds, are certainly descended from the ancient Celtic swifthounds – *Canes celeres* or war dogs.

How the Celts came into possession of these hounds can only be a matter of conjecture and is linked with the confused issues of early Celtic history. The Celts, like the Aryans before them, appear to have erupted from areas north of the Alps during the seventh and sixth centuries BC. They spread westwards over southern Germany and Gaul into Britain and Ireland and southwards into Catalonia; eastwards, they advanced and settled over the Hungarian plain and into areas as far removed as Anatolia. Over these regions they appear to have developed a tenuous dominion with trade and other links which existed until Roman times. Probably their expansion was facilitated by the possession of iron weapons, when their opponents had only bronze. In later times they were renowned for the fact that they introduced chariots and horses to warfare in western Europe and used massed dogs or hounds, presumably in association with the chariot charge, against their enemies.

The use of hounds for this purpose was not original, because mastiffs were widely used as war dogs in the Mediterranean world. The use of dogs of greyhound type, however, appears to be unique, though apparently effective. For instance, when the Celts invaded Greece in 273 BC their success is attributed partly to these dogs. Even earlier, in 391 BC, the Roman Consul Quintus Aurelius Symmachus received a gift of seven Irish wolfhounds which were sent to Rome for the Circus. The incident is described by Jennison [29] in his well-known book, *Animals for Show and Pleasure in Ancient Rome*. Apparently this Roman noble made long and expensive preparations for his *ludi*, but there was a scarcity of wild animals and he was reduced to showing Irish wolfhounds and crocodiles. The dogs, however, caused as much excitement in Rome as if unknown wild animals had been shown.

These dogs were named *Scottici canes*, but at that time the Scotti lived in Ireland. It is they who are supposed to have taken the Irish wolfhounds to Scotland to form the ancestors of the Scottish deerhounds. Although there is little authentic information about the origin of the Scottish deerhounds, it may be taken as certain that they were derived from the Irish wolfhounds. They were described by Queen Elizabeth's physician, Dr Caius, in 1573, and were well-known in the British Isles in the sixteenth century, when they were called the wire-haired British greyhounds.

Some authorities suppose that the Irish wolfhounds were taken to Ireland by the Phoenicians when they traded there in the last centuries BC. This, too, indicates a possible origin from Seleucia and thus a close link with the salukis. An origin from the known Celtic dogs appears more probable and their shaggy appearance may suggest greater affinities with the Persian salukis, the borzois, or the Afghan hounds. The Celtic dogs could easily have been acquired from these sources at the time of the Celtic expansion.

Another breed of greyhound is the Portuguese podengo. It was derived from North African greyhounds taken to Portugal by the Carthaginians and subsequently by the Moors. These greyhounds were crossed with local sheepdogs, but kept their greyhound characteristics though forming a distinct breed.

Greyhounds have been crossed with many other breeds. In some cases the aim was to introduce greater speed without loss of the original character-istics and in such cases the greyhound blood is so diluted that it is not apparent. The well-known lurcher was derived by crossing greyhound with retriever. The whippet, which from its appearance is a typical grey-hound, is a 'constructed' breed of about 100 years old. It probably origi-nated from the miniature Italian greyhounds which were too small for coursing, and various terriers.

The miniature or toy representative of this group, the Italian greyhound, is a true greyhound. This beautiful and fascinating little animal was developed by selection from very small greyhounds in response to the demands of fashion in the sixteenth century. It has remained deservedly popular until the present day.

Throughout the centuries, even millennia, the greyhound group, like the dingo, has retained certain basic characteristics and virtually varied only in respect of size and nature of coat. The nature of the coat has varied – as with all mammals – in response to different climatic conditions. As with the dingos, the lack of variability may argue that the greyhound group of dogs has been derived from a single ancestral form of wolf; the most probable conclusion is that they are independently descended from some cursorial form allied to *C. l. arabs* but now extinct.

A possible clue may lie in the records of ancient Egypt, as quoted by Loisel [39] that for hunting the Egyptians used dogs, cats, wolves, hyaena-like dogs, real hyaenas, leopards or cheetahs, and even lions. He states that they used wolves on the plains, but discontinued this in the Twelfth

Dynasty. The dogs used were (1) the fox/dog cross; (2) the dongolah dog, probably the pariah; (3) the great running dog, obviously another prototype greyhound, possibly the saluki. The hyaena dog possibly suggests the animal known today by the same name or the Cape hunting dog, *Lycaon pictus*. The reference to the use of wolves on the plains is fascinating and would appear to support the theory that there were wolves adapted to plains (cursorial) hunting conditions. The fact that the practice was discontinued in the Twelfth Dynasty might suggest either that these wolves had become scarce or disappeared because of changes in climate, or else that they were not worth the trouble after the ancient Egyptians had developed two cursorial breeds of dog.

Chapter 8

The Mastiff Group

WITH the mastiffs, we have a group of extreme diversity and great antiquity. Indeed it is not at all obvious that the various breeds grouped together here belong in the same category; and although points of similarity indicate that they should be discussed together, they are probably derived from multiple ancestries and this would account for their variability.

The breeds considered here all came originally from the same geographical region, the great mountain spine stretching from east to west across Asia and Europe, comprising the Himalayas and the mountains of Tibet, the mountain regions of Anatolia, the Alps, the Massif Central of France, and the Pyrenees. In these regions were developed breeds of dogs which, despite their diversity of form, size and colour, resemble each other in the acuteness of their powers of scent, the possession of a pronounced 'stop', a tendency to produce fine, silky coats and large floppy ears, and to have short or very short muzzles.

The wolves ancestral to the larger breeds at any rate are likely to be the woolly Tibetan wolves or similar varieties existing in these mountain ranges. Some breeds, however, may well have been derived in part from Asian wolves in northern India, and there is evidence that some breeds also contain both spitz and pariah blood. It is, therefore, not surprising that the derived breeds should be more variable than those of the dingo and greyhound groups and even than the northern group of dogs.

The mastiff group comprises the mastiffs, bulldogs and related breeds, such as the Molossian, Hyrcanian, Pyrenean mountain dogs, great dane, St Bernard, Newfoundland, and other very large dogs; the true scent-hunting hounds of variable size and form from the bloodhound to the dachshund; dogs of pointer type; the retrievers; the spaniels and setters, comprising the 'water-spaniel' and poodle group, the 'land spaniels' and setters, and the miniature spaniels; and lastly, small dogs of pug and pekingese type.

Many of these breeds are so mixed together with those of other groups that it is difficult to classify them with any degree of certainty. Furthermore, breeds with similar traits appear to have been developed independently in more than one area, so that superficial resemblances may to some extent be misleading. Nevertheless, where canine breeds are crossed, usually one breed leaves a greater imprint than another so that there is some hope of introducing a degree of order to this very confusing picture.

Mastiffs were probably first brought into domestication as shepherd dogs, for the purpose of protecting sheep flocks from the depredations of wolves and robbers. Throughout ancient times, fierce shepherd dogs of mastiff type were evidently widely distributed throughout the mountainous areas where sheep were kept. This would account for the reputation of these breeds for courage and fierceness, and indeed for the fact that the earlier breeds were accounted untrustworthy; most of the modern breeds derived from them, however, are companionable, extrovert creatures, not particularly exclusive in bestowing their affections.

In the ancient world, gigantic dogs of mastiff type were, as we have seen, owned by the all-conquering Assyrians and used by their kings Tiglath-Pileser and Ashur-banipal both in war and for hunting large animals such as lions, elephants, and wild horses.

In Greece and Rome, the Molossian dogs were held in esteem only second to the Hyrcanian dogs from India. The Roman legionaries introduced Molossian mastiffs into the mountains of the Iberian peninsula and in this way these dogs came to share in the ancestry of the Pyrenean sheep (or mountain) dogs. The Pyrenean mountain dogs are now a recognized breed of very large and exceptionally handsome dogs. They have similarities to the land spaniels, which are wholly or partly derived from them: this is one of the definite links that can be recognized in this confusing group.

The mastiff breeds survive also in India, as we have seen, both in the Nepalese mastiff and the bunjara which is probably similar to the ancient Hyrcanian. The Hyrcanian peoples lived in Baluchistan and northern India, in areas now incorporated into Persia and Pakistan. According to legend, the original Molossian mastiffs were taken to Hellas at the time of the Graeco-Persian wars, so that the Molossian and Hyrcanian breeds in all probability shared a common ancestry. From the same source were derived the mastiffs, which were ancestral or partly ancestral to the very large Hungarian komondor and kuvasz breeds, discussed in chapter 6.

These breeds may be included in the southern group of mastiffs. There is evidence that a more northerly group of mastiffs was established also in very early times as a result of different migration routes. The original

Tibetan mastiffs appear to have been somewhat heterogeneous, though there is not enough information to make it possible to describe them. However, Professor H. Kraemer of Berne, who has made extensive studies of the prehistoric dogs of Switzerland, states that the Tibetan mastiffs, indigenous in their native highlands for 2,500 to 3,000 years, are of two or three different types. One of them closely resembles the St Bernard and, according to Professor Kraemer, a large type of dog resembling this Tibetan mastiff and the St Bernard was present in Swiss areas in late Neolithic times. Skulls and skeletons of such dogs have been recovered particularly in the region of Vindonissa. He points, among other resemblances, to the existence of double dew-claws, also found in Pyrenean mountain dogs.

The original St Bernard could well have been derived from dogs of this type and in the light of Professor Kraemer's studies it is possible to postulate a plausible ancestry for the northern group of mastiffs. One such is the breed now known as the St Bernard, developed in the monastery of St Bernard de Menthon, which was founded in 692 AD. At this monastery, the monks ministered to the needs of travellers and, with the aid of their dogs, rescued those who were lost in snow drifts. These dogs have a remarkable sense of direction and ability to predict avalanches. They were derived from the local sennen hounds, known as *Berner-sennenhund*. It appears, however, that some southern mastiff blood was introduced into this breed in Roman times, when the legionaries brought Molossians to help them guard the Alpine passes; these were crossed with local sheepdogs and, it is stated, with German mastiffs.

The original St Bernard strain became weakened by excessive inbreeding and in 1830, in order to improve its size and strength, it was crossed with two other mastiff breeds, the Newfoundland and the great dane. The original strain was eventually eliminated by some epidemic disease, probably distemper, in quite recent times, and a new strain was developed by crossing Pyrenean dogs with wolfhounds. The modern breed is thus not the original one, nor is it pure mastiff.

We have seen earlier that, whereas large dogs of greyhound type were adopted by the Celts, the Germanic tribes possessed mostly dogs of mastiff type. Since the Germanic peoples are supposed to be predominantly of Alpine race, this is readily intelligible in the light of Professor Kraemer's discoveries. Mastiffs were supposed to have been first brought to Britain by the Angles and Saxons. They would thus be related to the German boarhounds or great dane, of which more below. The enormous French mastiff known as the dogue de Bordeaux is traditionally descended from Molossians and no doubt the Romans did bring Molossian dogs with them to Gaul. According to another account, they were brought to Guyenne by oriental peoples known as the Alains, supposed to represent the ancient Alicantes. This interpretation could well be a somewhat unintelligent guess, since the name 'Alain' could just as well be a corruption of 'Allemanni', peoples of German origin who invaded France at intervals between the fall of the Roman empire and the time of Charlemagne.

The latter interpretation is the more probable since in both France and Britain, mastiffs were known until the Middle Ages as 'Alains' or 'Alaunts' and a separate derivation is, therefore, improbable. In Britain, they were also known as 'bandogs' because of the necessity to restrain them with a band or rope. In this country mastiffs were used originally to hunt bears and wolves, both of which were to be found in Saxon times; in the Middle Ages they were used in the sport of dog fighting.

The Newfoundland dog, also a mastiff, was introduced to Newfoundland from the Pyrenees by Basque fishermen and developed as an independent breed there. This view, however, is disputed by Ash, who states that it was introduced by Norwegians in the sixteenth and seventeenth centuries and. that it was developed by selection and crossing with other introduced strains. Ash states that the Norwegian peasants kept hunting dogs resembling the Newfoundland, which they used in mountainous areas for hunting bears and wolves. The Newfoundland may, therefore, be derived from crossing northern and southern mastiffs derived from Norway and from the Pyrenees.

The great dane is the subject of dispute between the Germans and the Danes. To the Germans, it is the German boarhound; to the Danes, the great dane. However, there can be little doubt that it is representative of the mastiffs kept in ancient times by the Germanic tribes, though it is supposed to have been crossed latterly with greyhound blood.

The English bulldogs were developed by special breeding of mastiff types in England during the Middle Ages, when they were known as Molossians. They are supposed to have been bred at first as butchers' dogs in order to control savage bulls with which butchers had to deal. It was only later that they were used for baiting and fighting bulls, when the short legs and powerful, low chassis was of value because the dog, being low to the ground, would be less vulnerable to the bull's horns.

For centuries, bull baiting was a British national sport and together with the characteristics of overshot jaws, great girth and low build, these dogs came to be endowed with ferocity, truculence, tenacity, and great singleness of purpose. Their means of attack on the bull was to seize it by the nose when it attempted to gore them with its horns. The incisor teeth, overlapping scissor-like because of the overshot jaw, would lock on this sensitive portion of the bull's anatomy; once locked, nothing could prize them apart, except the pouring of water down the bulldog's nose. It is recorded that on one occasion a butcher had matched his elderly female bulldog against a bull. Once locked on the nose, in response to a bet he amputated the dog's fore and hind legs above the feet but failed to induce it to release its hold. He then slashed the body in two with a sickle, but this too failed to dislodge it and in death the teeth remained locked on the bull's nose.

Strangely enough, a similar tale is told of an episode which took place in classical times with a Hyrcanian dog in India. In this case the dog had its jaws locked on a bull's nose and the owner amputated both hind legs without dislodging it; he then severed the head from the body and in death the teeth remained fast in the nasal tissues of the bull. The resemblance between these stories suggests that among the Hyrcanian dogs there may have been developed a breed with qualities resembling those of the early English bulldog.

It was around the beginning of the present century that the English bulldog was bred as a house dog rather than for bull baiting. It has taken many years to eliminate the ferocious instincts which had previously been developed and which embodied the contradictory characteristics of the mastiff breeds: savagery and ferocity, combined with tremendous courage and tenacity. Today's bulldog is a docile and companionable creature, though he still retains the appearance of his pugnacious ancestors.

The smaller French bulldog was developed in France by crossing English bulldogs with small litter mates of bulldogs from Spain.

The boxers are another breed derived from the mastiffs although their origin is somewhat controversial. The *Prisma Encyclopédie* [51] regards them as derived directly from the old Molossian war dogs, imported into Britian in the Middle Ages. This authority denies that there is bulldog in their ancestry, though other authorities believe that the boxer is derived from crossing Germanic mastiffs with English bulldogs. Such a cross is known in the bull mastiff, which was developed by crossing the English mastiff and fighting bulldog in the 1850s. This dog has the courage and strength of the bulldog with superior speed and was extensively used for the protection of large estates against poachers.

Finally, we have already noticed that the bull terrier was developed by crossing bulldogs with fox terriers in order to produce an animal with the tenacity of the bulldog and the speed of the terrier. This breed, unlike most that are mixed, appears to be almost intermediate between the two breeds, perhaps tending rather more towards the terrier type.

In pursuing the enquiry into breeds derived, or mostly derived, from mastiff types, it is necessary to study the true hounds. Among the mastiffs we have found dogs with the worst and the best traits inherent in canine breeds. In the hounds, spaniels, and setters, whatever their ancestry, we encounter breeds in which man by selection and crossing has developed very specialized properties for various forms of hunting. The mastiff strain has been able to produce these properties more successfully than other strains. The nearest link between true mastiff and hound is possibly the bloodhound, and it is interesting to note that one variety of Tibetan mastiff closely resembles the bloodhound in head and shape; this variety is generally supposed to be ancestral to the Molossus, and possibly the hounds have been developed independently from the same strain of Molossian.

The true scent-hunting hounds are thus probably descended from some ancestral dog of mastiff type resembling the bloodhound. Rawdon B. Lee [37], in his first edition of *Modern Dogs*, states his belief that no modern breed of dog is so like his progenitor of more than 3,000 years ago as the bloodhound. Hounds resembling the bloodhound were probably developed parallel to the Molossian type of mastiff and used for thousands of years for hunting. They have in acutely developed form the mastiff's fantastic powers of scent.

The modern breed of bloodhound is descended from the black St Hubert's hounds, kept in the sixth century at the Monastery of St Hubert in the Ardennes. They were imported into Britain, where they found a ready use in hunting down outlaws and criminals. They were used by both English and Scots in border forays to follow raiders, who much enjoyed the sport of slipping over the frontier for games of murder, rape, and cattle theft. The bloodhounds were greatly feared and left a tradition of awe and horror undeserved by so amiable a creature.

Another hound of ancient lineage closely allied to the bloodhound is the Weimaraner, derived from the old Schweizhund, a solid red bloodhound with powers of scent said to be equal to those of the bloodhounds.

Modern hounds are all derived from some early bloodhound type. They vary in size, shape, length and shape of leg, and colour. All have been developed by selection for special purposes, but nevertheless remain variations on a single theme. They all have wonderful powers of scent, will combine to hunt in packs, and have rather long floppy ears and silky coats.

The smallest of the hounds is the German dachshund, derived from early German hounds, the Deutsche Brachen. These little dogs were developed as hunting dogs for use in water and also to dig into burrows. Today they have been bred into miniature breeds and make the most delightful pets. The legs are bowed and the feet out-turning, probably due to deliberate breeding since this helped a short-legged dog in digging. The short misshapen legs are sometimes attributed to a genetic mutation and ascribed to an abnormality known as chondro-dystrophia.

Other breeds of short-legged hounds are the drevers, the Danish drever or Strelluftswer and the Swedish drever. Both these breeds originated in Germany from the original German hound, the Westfalische Dachsbrachen, thus having a common ancestry with the German dachshunds. Swedish dogs were also mated with the Swiss hounds.

The Swiss had two breeds of foxhounds, both of which have the same origin, the Swiss Laufhund and the Luzern Laufhund. These are believed to be descended from foxhounds of the Celtic peoples and to have lived in the same homeland in Switzerland at least since the fifteenth century.

The English hounds range widely in size, developed for different purposes of hunting, but they do not seem to have become prominent in England until the twelfth or thirteenth centuries. The traditional English hound, said to have been present in these islands since the days of the ancient Britons, is the beagle. A still smaller breed, the bassett, was introduced from France during the nineteenth century and became popular. The first known pack of English foxhounds was owned by King Edward I in 1299 and was derived from a mixture of hound blood, probably with bloodhound predominant, and selective breeding. Harriers (vide the greyhound group) were developed in England around 1260 and otter hounds in King Edward II's days around 1310. The latter was rather a rough type of dog, probably produced by crossing small hounds with terriers.

The true hounds are thus a specialized group of scent-hunting animals, developed during the time when the larger types of wild fauna were disappearing and there was the need to control smaller pests such as otters and foxes. Their ability to provide sport in the chase led to the great popularity of this kind of hunting which, in spite of opposition and criticism, persists to the present day.

The second great group of sporting dogs derived predominantly from mastiff ancestry is that of the pointers, spaniels, setters and retrievers. These dogs, too, are mostly products of man's ingenuity in breeding dogs with special characteristics of value for hunting and similar sports.

Dogs of pointer type are of very ancient lineage. Today they are distributed throughout mountain areas between northern India and western Europe. Such are the dogs of the gypsy caravans and the dalmatians – one distinctive breed of pointer – supposed to have been brought with the gypsies from India to Dalmatia, whence they reached Britain in the eighteenth century.

We have already suggested that Neolithic dogs, such as the Windmill Hill type, were of this kind and were used as general purpose hunting and herding dogs. If this is correct, they date from at least 3000 BC and may have been one of the breeds which were brought during Neolithic migrations from the Black Sea area. From their conformation, one would suppose a partial origin from early dingo (as opposed to pariah) stock; if this is so, an origin in northern India is probable. A feature of dogs of this type is their multiple markings, with flecks and spots on coats of different colours, qualities which they have imparted to the spaniels with which they are evidently mixed. Welsh sheepdogs are typically of this type.

The pointers came into their own when a suitable dog was required in early times for hunting with hawks. They were later in demand when sportsmen shot birds with muzzle-loading fowling pieces as described in chapter 10. Their quiet movements and slow, deliberate action made them especially useful for these purposes.

It is said that retrievers were bred in recent times from spaniels. This can hardly be so because the oldest known retrievers in western Europe, the braques français and the braques d'Auvergne, were present in France long before the development of modern spaniels. Retrievers have the typical prominent 'stop', head and ear formation, powers of scent and temperament of dogs of the mastiff group. They are also good water dogs, and it is possible that the original curly-coated retrievers were crossed with primitive water-spaniels; it is also probable that they were mixed with pointer blood.

The retrievers mostly used today are the labradors, which were derived from dogs taken to Labrador in the seventeenth century by the Breton cod fishermen and therefore probably derived from the French braques. They have been intensively bred for retrieving purposes in Britain since their reintroduction to Europe. Originally they were black, but golden and other colour variations are now available.

Traditional theory would derive all western spaniels from small varieties of Pyrenean mountain dogs. The name 'spaniel' would support a supposed origin from the French 'espagnol', and probably one line of ancestry is derived from this source. However, this cannot possibly be the whole story, and it is necessary to revert to the ancient distinction between the 'water' and 'land' spaniels to understand the position. These two types were originally very dissimilar.

The earliest 'water' spaniel is represented by the French barbet, which is thought to be ancestral to the 'water' spaniels proper and to the poodles. The early poodles were much bigger dogs than they are today and were extensively used in the hunting and trapping of water fowl. They had rather long, rough coats, which became heavy with mud and water, so that they were traditionally clipped in rather the same way as the miniature poodles of today. They were very similar to the 'olde English water-dogge', supposed to be ancestral to the English and Irish water spaniels. The olde English water dogge closely resembled the French barbet and in old days used to be sheared from the 'nauille downeward or backeward'.

The early 'land' spaniels are quite different from these animals, although in the more modern breeds they are probably to some extent crossed. The true spaniels comprise the Breton spaniels – a long-legged variety – the German spaniels (Wachtelhund), Sussex spaniels, English springers, Welsh springers, cocker spaniels, and more recent breeds such as the clumbers, field, and so-called American water spaniels. The Breton and similar spaniels from other parts of France with their long legs appear to be the most primitive, and could well be derived from small Pyrenean mountain dogs (of which they have the head and ear conformation) and pointers, from which they derive their coat colours and markings.

Both springer and cocker spaniels were developed in England by deliberate selection in Romano-British times. The springers were used to 'spring' game and the 'cockers' were a smaller breed for flushing woodcock. In early days they were used with the hawk, but in later times until the present day were useful general-purpose dogs with the gun, capable of both flushing and retrieving game. The bigger spaniels are also good water dogs and will retrieve large birds, such as geese, from water or marsh.

Setters were deliberately bred from spaniel stock when a faster moving dog than the pointer was required for use with breech-loading guns. The setter will 'sit' or 'set' once aware of the presence of game in the underbrush and will then, at the command of his master, flush the bird and retrieve it when shot, without the same maddening deliberation of the pointer which was so useful in the days of the muzzle-loader.

According to the Hon. Mrs Neville Lytton [42], the miniature spaniels of the so-called King Charles and Blenheim types owed their origins to importations of toy spaniel breeds from the Far East to Italy in mediaeval times. They would thus be most closely related to the Tibetan spaniels, Japanese spaniels and pekingese, and of different stock from the European breeds of spaniel. Mrs Lytton points out that at that time miniature dogs were much in demand in upper-class circles in Italy and that the miniature Italian greyhounds were also developed then. She believes that the King Charles was developed by crossing the short-nosed miniature breeds from China with Maltese dogs.

These little spaniels were, of course, notoriously the favourites of King Charles II, after whom they are named. They were certainly not introduced to England by him since, among others, his father King Charles I owned one. Earlier than that, they were known in the days of King Henry VIII and were described by Dr Caius, Queen Elizabeth I's physician, as 'delicate, neate, and pretty kinde of dogges'. Both the cavalier King Charles and Blenheim spaniels were derived from the King Charles.

Other small spaniel breeds of Far Eastern origin are the French papillon, supposed for more than 400 years to be French, but introduced by way of Spain from the Far East. The griffons, the short-haired brabançon, and the long-haired bruxellois, are small dogs of spaniel type probably of Far Eastern origin, developed during the fourteenth and fifteenth centuries. The Japanese spaniels originated in China around 500 BC and came to Europe by way of Japan. They reached Japan as the result of a fine pair given to the Mikado by an emperor of China and became very popular there.

The Far Eastern spaniel-type dogs were very likely developed from the smaller Tibetan house dogs, such as the Lhasa Apso and the rather larger Shi Tzu, which we have already met. In addition to these dogs of spaniel type, the Chinese 'ha pa' dogs were imported into Europe, where they were ancestral to the 'pug dogs' and probably in part also to other breeds of short-nosed dogs such as the griffons. Pugs became especially popular in Holland and one was the inseparable companion of William of Orange. However unlikely it may seem, these little dogs were derived from small members of the original mastiff stock which gave rise to the large mastiffs, such as the Molossian and great dane.

Little more need be said of the pekingese, which has already been described. It is a specialized breed of the Chinese 'ha pa', perhaps crossed in very early times with Maltese.

Dogs of the mastiff group are the most diverse and probably also the most popular of all types. Their origins and breeding have been so confused that it is difficult to give a coherent account of them. Nevertheless, the various members are linked by recognizable characters, and gradations between the different forms can be traced. The extreme variations encountered suggests multiple ancestry and crossing with dogs of different origins. The original home of the breeds was probably Tibet which, in spite of its present remoteness, was in early times a centre of trade and a meeting place between east and west and north and south.

Part Three

The Role of Dogs in Human Affairs

Chapter 9

The Uses of Dogs

WE have now traced the history of dogs from the parent wolves through various stages of domestication to the present time. It is a strange story and one would dearly like to pry into the past and share the lives of our ancestors of 20,000 years ago – in relative terms no great age – to see how they lived and learn in detail the processes which led up to the development of the many specialized breeds of dogs that exist today.

It is evident that in the early days of domestication, the dog was just another tool, acquired by man in desperate circumstances. It was acquired for its usefulness and a study of the uses to which it was put must be made to complete the story. Certainly the original purposes for which the dog was used were purely materialistic and a large section of the canine population – the pariahs – was discarded and despised when of no more use. The developing sentimental attachment between man and dog has complicated an originally simple picture. Therefore, in studying the uses of dogs, it is necessary to confine ourselves to those which have influenced their breeding; it would be inappropriate to discuss *all* the various purposes for which dogs have been used.

In early days – and today in arctic regions – dogs were used for rather simple purposes and for basic needs. They were used for drawing sledges and other forms of transport, for hunting, herding, and guarding flocks. They scavenged, and in turn their flesh was used for food and their skins for clothing. There is little doubt that among some races of men at any rate, dogs were involved in mystical ceremonies and magic and were used as sacrifices to the gods. As men came to live in permanent settlements, dogs acquired many further responsibilities, such as guard duties, war and police work, the tracking of raiders and lost persons, and the control of pests such as rats and mice.

These uses have all dictated the selection of dogs with particular properties and thus contributed to the production of different breeds. Dogs have also been bred for debased purposes – for fighting other dogs and wild animals, and even humans in circuses, and for bull-fighting – and they have been bred for more noble objects, as companions and pets. As such, they have often been portrayed by artists in painting and sculpture. From time immemorial, dogs have been important in trade and commerce. Today perhaps the greatest single incentive in the breeding of dogs is to acquire perfection for purposes of showing and ultimately for profitable sale of the offspring.

At the beginning of our story, the northern peoples were well clothed in the skins of animals they killed; evidently they used skins for many other purposes as well, perhaps for tents and for simple sledges to drag their possessions over the snow. As draught animals, dogs can carry only light loads; but they can pull heavy ones. It is possible that even in ice age times the travoise had been invented; this is a device, still used by American Indians, which is hauled by dogs. It is made from two sticks arranged in a V; the junction rests on the dog's back, while the two ends trail on the ground behind.

During Mesolithic times, men certainly travelled on skis and they had well-constructed sledges with wooden runners, not unlike those of today. There is a fascinating picture reproduced by Graham Clark and Stuart Piggott [11] of men on skis, which recaptures something of the glow and glamour of the snowscape; in spite of the crudeness of the drawing, the figures show an exuberance and love of a life which was to be so ruthlessly and drastically changed. The transformation man endured has been described in moving terms by these authors in their revealing book *Prehistoric Societies*.

'We take leave of the Advanced Palaeolithic hunters of western Europe when at the peak of their attainment their whole world was about to disintegrate with brutal suddenness before the impact of environmental change. In their art, applied to small objects and to the walls and roofs of their caves and shelters alike, they left behind a legacy widely recognized as one of the supreme achievements of mankind, superior in significance to the more parochial attainments of the civilized peoples of antiquity, because more universal in its relevance. Both conceptually and physically the Advanced Palaeolithic peoples were true representatives of modern man, representatives who tried out as it were for the first time the faculties by which during the astonishingly brief period of 10,000 years were shaped all the diversities and intricacies of civilization itself.'

Strange indeed that these people, in generations so close to us, possessed the capacity for achievement and invention on this tremendous scale, an advance comparable in its way with latter day achievements of science. It all started with the recruitment of wolves, the domestication of the first animal; without this the winning of an adequate livelihood would have been accompanied by great difficulties, and advances would inevitably have been achieved at a slower pace. This is as strange as the story of man himself during this extraordinary period, a success story to which the dog himself has contributed by his own characters and virtues as they were developed by man.

An outstanding example of the way in which dogs have helped man to overcome almost insuperable difficulties are the polar regions. Until recently, when vehicles adapted to snow conditions have been developed, Arctic and Antarctic exploration would have been impossible without dogs to draw the sledges. At first they were close to the parent wolf stock and methods of treating them were harsh and brutal, involving the imposition of the master's will on a savage wild animal. As already mentioned, these methods have been described by polar explorers who, by proper use of the dogs and by humane treatment, are able to obtain better results than the Eskimos.

The huskies of Greenland and Alaska are a heterogeneous group, very close to the parent stock because of back-breeding to wild wolves. Little attempt has been made to breed selectively from dogs which have special aptitudes for work. Certainly they respond to kind treatment and better results can be obtained in this way than by the harsh traditional methods.

The related samoyeds have been bred by the Samoyed tribes of northern Siberia for sledge work for many centuries, with the result that they are more teachable, docile, obedient and efficient. Nevertheless, many polar explorers put their faith in the huskies because of their great hardihood, resistance to cold, and physical endurance.

Dogs have been important as draught animals in temperate countries until comparatively recent times. They were used for pulling wheeled vehicles and to this day a two-wheeled driving cart with cross seats back to back is known as a dogcart. They are not often seen nowadays, except occasionally in country districts and when petrol restrictions drive motorized vehicles off the roads. However, some will remember from their younger days the exhilaration of driving a dogcart along a country lane with a spirited pony between the shafts. The use of dogs in these vehicles was prohibited by law in Britain in 1885, so that there can be few if any still alive who remember the original draught animal. In Belgium, however, dogs were used very extensively in this way and as late as 1939, earnings from dog haulage were estimated to be £2,000,000.

In the eighteenth century, the dalmatian was brought to Britain and quickly became a great favourite as a carriage dog. The dalmatian, with specialized groupings of black spots on a white coat but otherwise of typical ancestral pointer type, was apparently taken to Dalmatia by bands of migrant gypsies; possibly because of this association, they developed an attachment for horses. At the time they were brought to Britain they had acquired such an affinity with this very dissimilar perissodactyl mammal that they were never happier than when allowed to bed down for the night in the horses' stable. It seems that the horses reciprocated this attachment and, although worried by the heel-biting proclivities of the traditional English terriers, they always tolerated and showed affection for dalmatians. Whether the gypsies deliberately bred dogs which would associate happily with horses is not known, but obviously it would be to their advantage if dog and horse were at least compatible.

Our eighteenth century forebears discovered that a dalmatian running between the horses added a touch of style to an equipage in which they took so much pride. Today, unfortunately, such sights are never seen, but one can imagine the superb display made by a four-in-hand drawing a nobleman's carriage with its magnificent horses, liveried coachmen, the crested vehicle, and the dalmatians running immaculately aligned between the horses.

Today the dog has little significance as a draught animal, but is still used for carrying collecting boxes for charitable purposes, for the carriage of messages in war, or for carrying his master's newspapers from shop to home. The dog is a willing ally in activities such as these and it is touching to see his evident pleasure when he masters some service by which he can be of assistance.

An important role of the dog in modern life is in the care of the blind. The uncanny instinct by which trained dogs lead their blind masters to their destination through crowded streets assuredly gives the lie to those who assert that dogs are not endowed with faculties of intelligent thought. As with hunting and other activities, a dog's success in undertaking such tasks depends both on his inherent abilities and on well directed and intelligent schooling.

Among western people today, the idea of using dogs for food is repugnant and those who do eat dogs are regarded as uncouth and primitive. But to some peoples the dog forms a tasty dish and is thought to be a useful and legitimate article of diet. Among the Samoyeds, the dog is used both for drawing sledges and for eating, while the fur makes warm clothing. The allied chows of China were probably derived from the samoyeds and, no doubt because of their influence, were regularly fattened and eaten. It is said that dogs intended for butchering are fed only on vegetable foods and that their meat is succulent and tasty. Until recently, dog flesh was sold in butchers' shops, in spite of a law prohibiting this in 1915. Even in China, however, there was some prejudice against this practice, because dogs were said to represent the guardian spirits of the houses and many Chinese objected to their being used as articles of food.

In Greek and even more so in Roman times, dogs were eaten regularly and dog stews were particularly relished. Although not eaten in Britain during the Middle Ages, dogs were used for food on the continent and English travellers commented without disgust on the appearance and taste of dog flesh. In Africa today, although the eating of dogs is not a universal practice, in many countries dogs are fattened for food and are eaten quite regularly.

In general, the eating of dogs seems to have been incidental, with the sole exceptions of the chow in China and the small breeds of Mexico and Peru. The chow was the only dog in China which was specially bred and fattened for eating and none of the other dogs present in that country found its way into the butchers' shops.

Little more need be said about the development of sheep- and shepherd dogs. It has already been shown that the potentialities in dogs for the control of sheep, cattle and reindeer was one of the driving forces in the selection of certain breeds, resulting in the great group of sheep dogs which comprises those of collie and alsatian types.

Dogs have been used for herding and protection for at least 4,000 years and famous breeds of ancient times, such as the Molossian and Hyrcanian, were primarily shepherd dogs which were later developed for other purposes such as for war and for police and rescue work.

The patience and cleverness of dogs in controlling animals is something which impresses everyone, even today. These dogs will respond to hand signals, voice or whistle and, owing to the greater range of the dog's ear, they can hear sounds outside the human auditory range made by supersonic whistles.

The use of dogs for keeping watch and for guarding premises whether public or private is as old as domestication itself. In western countries in recent years dogs have been bred to be more docile and friendly; but many will remember the noisy, yapping dogs of a generation or more ago which were kept by poorer people to watch over their possessions. Such dogs were often tied near the entrance to the house with a lead sufficiently long to permit them to get a hearty bite at the heels of anybody approaching the front door when the owner was not present. This was certainly an effective way of keeping out unwelcome intruders such as policemen and tax inspectors.

Today, with crime on the increase, people are again keeping dogs in their flats and houses to deter intruders with felonious intent. For these purposes dogs should be trained – as they easily can be – to accept food from nobody other than the owner, so that they cannot be doped or poisoned by the intruders. The function of dogs in property protection is twofold: first, they must bark and arouse the occupants and secondly they should be fierce and trained to attack intruders. Many modern breeds have been selected so much for amiability that they are of little use as watchdogs and merely wag their tail and make friends with the thieves. Obviously there is a case here for breeding again strains of noisy, bad-tempered dogs.

In the past, the dogs used mostly in police work were bloodhounds, which were especially selected for this type of work. One of their main features is their fantastic sense of smell, which enables them to follow a particular criminal for many miles over difficult country, distinguishing one individual from all others even if the scent is two days old.

Dogs of bloodhound type have been present in mountainous areas of Asia and Europe from time immemorial, though the modern breed is believed to have been developed by the crossing of several different strains: the St Hubert, talbot, and the old southern hound. St Hubert was the son of the Duc de Guienne (656–727 AD) and is supposed to have been miraculously converted to Christianity in 708 AD as a result of seeing a stag with a cross between its antlers when he was hunting on a Sunday. He later became Bishop of Liège and founded the monastery that still bears his name. The abbots of this monastery have always bred hounds of the same strain, which were originally black. The talbot hound was introduced into England by William the Conqueror and is believed to be a descendant of the Hubert breed.

Many stories could be told of the use of these hounds for the capture of slaves, criminals and other wanted persons. In particular, they were used in the border forays between England and Scotland before the Act of Union and were known as 'slough' dogs. When Scottish intruders attacked the English border towns, often killing and maiming the inhabitants, they were pursued by bloodhounds and were often cornered and apprehended. The virtue of the bloodhound was that it could select the true miscreant from among innocent persons. The unfortunate Duke of Monmouth was literally run to earth by these dogs after the battle of Sedgemoor; and Robert the Bruce narrowly evaded capture by wading through a stream and thus dispersing his scent.

Both on account of their unfortunate name and because of their sinister reputation, the use of bloodhounds was largely discontinued in the nineteenth century and they are no longer so extensively used by police forces. They acquired a particularly bad name from the so-called 'Cuban bloodhounds', which were certainly not of pure blood but were possibly half-bred foxhounds. In 1795, 600 maroons (freed negroes) revolted in Jamaica and it took a force of nearly 5,000 soldiers to track them down. Even so, they were not apprehended as they had hidden in inaccessible mountain bolt-holes. The government of Jamaica brought 100 of the so-called 'bloodhounds' from the nearby island of Cuba with twenty dog handlers. By these means the negroes were tracked down and forced to surrender and were later transported to America. A certain John Macleod raised the question of the morality of this action by the government of Jamaica in Parliament. There was great public concern and the government of Jamaica was ordered never to take such action again.

Today many breeds of dog are used in police work, boxers, alsatians[1] and many others. Most people have seen exhibitions of the work of these dogs and will appreciate the great cleverness which is displayed both by the dogs and their handlers. For successful police work each dog must be trained and tended by its own handler, for whom alone it will work.

Some mention of dogs in war in early times has been made in previous chapters and need not be greatly enlarged here. The immense mastiffs bred by the Assyrians, Persians, and Babylonians as shock troops, which hurled themselves in massed packs on the front line of the enemy, must have been, with their savagely spiked collars, an unpleasant ordeal. Apparently, however, they were not really effective, because their use in this way did not continue into Greek and Roman times; it may be supposed that these dogs were no match for well equipped infantrymen armed with long spears and swords. Nevertheless, it is recorded that the Celtic conquest of Greece was accomplished with the help of their traditional greyhounds.

[1] A breed specially produced in Germany for police work was the Dobermann Pinscher.

It seems that the Greeks had no knowledge of hounds of this size, possibly more of the nature of the Irish wolfhound than the modern greyhound, and were overawed by these savage beasts.

In mediaeval times dogs (possibly mastiffs) were given armour and were trained to enter into the fray during the course of armed equestrian combat and to bite the horses' legs – surely a most unsporting way of unseating an opponent in days of chivalry !

In all wars dogs have found multifarious uses, as for example in ambulance work and in finding people lost under debris. They have been used for sending messages; they are driven in front of advancing troops or jeeps to detect mines; they even enter into battle as mascots with the regiment, or as a companion for its commander. In armies, both at war and in peace, dogs are in demand for the same uses as in civilian life, namely for guarding premises such as arsenals and for patrolling areas to which entry is forbidden.

Dogs have distinguished themselves in many wars. A poodle known as 'Moustache', born at Falaise in Normandy in 1799, joined a regiment of French grenadiers and accompanied it through several campaigns, being present at the battles of Marengo and Austerlitz. At Austerlitz the ensign who carried the colours was killed, whereupon Moustache took them back to the regiment. Subsequently he joined a regiment of dragoons and died honourably of wounds in Spain.

During the civil war, Prince Rupert was invariably accompanied by his poodle 'Boy'. Boy took part in all his campaigns and died of wounds at the battle of Marston Moor. Both in peace and war, a pug went everywhere with King William III. The Duke of Marlborough always took his King Charles spaniel with him and it is depicted on tapestries at Blenheim Palace, where it is shown as being present at the battle of Blenheim itself.

In modern times, the uses of dogs in war are more specialized than in days of old – as when the Knights of Rhodes maintained the famous mastiffs who knew a Turk from a Christian by the smell, or when Spanish bloodhounds helped in the conquest of Mexico and Peru. In those days the term 'dogs of war' was one to strike terror, conjuring up visions of immense hounds three to four feet in height, of great ferocity and armed with broad collars with long iron spikes.

Today there is little breeding for purposes of war, although in early times the Assyrians and other warlike peoples bred them in great numbers especially for this purpose. Qualities of ferocity or docility and amiability can soon be bred into, or out of, the canine species and there was little difficulty in breeding dogs for bellicose purposes.

For the control of pests such as rats and mice the dogs *par excellence* were the terriers. Developed chiefly in Great Britain, they were encouraged also to go underground after foxes and badgers. In Australia, dogs have also been used with great effect in connection with rabbit extermination programmes, where their function is mainly to scare the rabbits so that they stay down the holes and in their runs while the pest control officers introduce smoke and poison. This operation is known in Australia as 'dogging'.

Dogs have been specially bred over the ages not only for uses in connection with man's economy, but also for his more sophisticated and aesthetic enjoyment. Small dogs such as the Maltese and pekingese, lap dogs, 'toy' dogs and 'sleeve' dogs were sometimes developed for special purposes in connection with certain types of hunting even for truffles, but also for pleasure and companionship. Small dogs can be bred from nearly all the ancestral groups; the tiniest dog in the world, the chihuahua, for instance, seems to have been derived from a terrier of black and tan type crossed with the small Mexican temple dog, which was already there at the time of Cortes.

It is not difficult to produce small dogs by selective breeding any more than it is difficult to produce or find small breeds of any other kind of animal. Nature herself adapts the size of animals to circumstances of life by means of natural selection, as witness the pigmy breeds of man, horses, cattle, sheep, goats, pigs, and poultry. Among the Canidae, *Canis lupus pallipes* is smaller than the northern wolves; the coyotes (*Canis latrans*), although in other respects very similar to other wolves, are smaller still. Among the domestic breeds of dogs, small varieties exist among all of those we have discussed so far. The miniature greyhound is that lovely creature, the Italian greyhound. The small spitz dogs are represented by the corgis and the Maltese dogs. Small terriers include the Yorkshire terrier and the black and tan; terriers range from this size through dogs such as the cairn and sealyham to the airedale. Among the short-nosed dogs there are the pekingese, King Charles spaniel, pugs, griffons and a great many others; as usual with the mastiff group, they show greater variability than the others. Even in the rather uniform dingo group, smaller breeds exist; there are the small New Guinea singing dogs, which may be seen at the London Zoo. These dogs, with their unusual kind of howl, are typical little dingos and have been bred, presumably selectively, by the natives in New Guinea.

The urge to produce small dogs to love and befriend has actuated mankind from Neolithic times to the present day. In places such as China they have come to be regarded as the guardian of the home and have thus acquired a spiritual or mystic significance. The earliest conscious effort to breed small dogs was probably made by Neolithic dwellers in lake settlements, who would need a smaller animal suitable to their rather confined dwelling quarters. Probably before the development of these small dogs, man's somewhat unwieldy canine friends would have been kept outside. Real affection and concern for the welfare of dogs would come with their admission to the home, where they would play with the children and keep the women company when their menfolk were away hunting, tilling or engaged on other duties. The Neolithic breeders seem to have gone about their task with considerable skill, if indeed their efforts resulted in the lovely little Maltese dogs with their silky coats and white colour, which were so popular from early Egyptian times until the Roman era and were in demand from western Europe to faraway China.

Much has already been said about the further development of breeds of toy and sleeve dogs, particularly in China and other far eastern regions, where the origin of the King Charles spaniels may lie. Some believe with Mrs Neville Lytton [42] that they were developed by wealthy Italians in mediaeval times from dogs imported from the Far East; others that they originated from Portuguese sources in the Far East and reached England with Queen Catherine Braganza in the year 1662 when she married King Charles II.

Once selective breeding began, it was inevitable that an element of competition should enter into it; even in classical times, as discussed in the first chapter, the merits of different breeds of dogs were disputed by Xenophon, Aristotle, Pliny, and other writers. However, the art of competitive breeding combined with showing is rather recent and has been intensified only since the Kennel Club became active in the nineteenth century. This has resulted in an improvement not only in the number of breeds, but also in the status of dog breeding and of dog breeders, formerly a disreputable class of persons.

Before this dogs were bred for fighting and matching against bulls and wild animals. In such activities, their breeders followed the ancient tradition of Roman times and earlier, when the ghastly spectacles of the arena were staged. Gladiator met gladiator in fair fight, but criminals and dissenters were torn to pieces by wild animals and dogs, debasing human nature itself. From the Middle Ages onwards, dogs were selectively bred for these purposes, for they needed to have the ferocity and conformation necessary to fight other dogs and to enable them to be pitted against bulls and wild animals. To these pursuits we owe the production of the bulldog from the ancient mastiff, as well as other breeds with overshot lower jaws; in them the front teeth lock like a trap, enabling the dog to get a hold on the bull's nose which could not be dislodged. The dog's nose was set back so that breathing was not impaired by being muffled in the animal's skin.

During the sixteenth and seventeenth centuries the bulldog came to be distinguished from the 'bandog', and was known as the Molossian – incorrectly. Camden who wrote at this period (around 1600) states that along the riverside from St Thomas' hospital was a row of houses among which 'is a place like a theatre for baiting bears and bulls with dogs and kennels of band dogs which are so strong and bite so hard that three of them are enough to seize a bear and four a lion'. This equation – 4 mastiffs = 1 lion – has been handed down and repeated so often that it has become the beginning and end of many a story about the mastiff. Both bull- and bear-baiting by dogs were barbarously cruel sports as the bear was blinded and the bull tormented; they lasted in England until the eighteenth century.

In the realm of sport, coursing dogs were developed for purposes of racing and were pitted against each other to pursue live hares, which were captured and released for the dogs to follow. This sport still survives, though greyhounds now race mostly in pursuit of electrically propelled instead of living hares. Dogs of this group are, of course, necessary for this sport because they are sight-hunters and will therefore follow the electric hare at great speed.

In the field of entertainment, dogs are popular on the stage, in the cinema and in circuses, and even the traditional Punch and Judy show survives in England. The latter probably came from Italy, though how the dog Toby came to be associated with these puppets is not easy to determine. Probably Toby represents the dog of Haji Avid, a mythical character in Turkish folklore.

The role of dogs in art and literature has been extensively studied and is described in many works, so that it need not be discussed in detail here. Pertinent to our argument, however, is the breeding of the pekingese in China for the sole object of producing an animal of superlative grace and beauty. In this instance the pursuit of artistic excellence may be said to have contributed to the production of a new breed and its varieties and so to the natural history of *Canis familiaris*.

In various parts of the world, dogs have been used in religious and magical rites. In ancient Egypt, dogs – together with almost all other animals – were held in veneration and in certain parts they were worshipped. Some people believe that the Jews' abomination of dogs was originally due to a recoil from the veneration accorded them by the Egyptians. Whether the Egyptians ever selectively bred dogs for magical purposes we shall probably never know. The development of special breeds of dogs for religious and mystic purposes in China and Mexico has already been discussed. Clearly there was an urge to produce small dogs as household guardians and from this arose the belief that, in the absence of a real dog, a porcelain or pottery substitute would do. Even in burial customs where the dog must accompany his master to the world of departed spirits, images of dogs were later accepted as satisfactory substitutes. The origin of these beliefs is extremely simple: the living dog keeps away corporate intruders and symbolically he protects the house and its owner from evil spirits.

Even the eating of dogs' flesh has symbolic significance among some peoples. Some American Indians ate dogs ceremonially to acquire courage. In the East Indies dogs' flesh used to be eaten to make the warrior bold and nimble in war. Such practices are usually preceded by ceremonial sacrifice of the dog. Among the bear-worshipping 'hairy Ainu', well-fed dogs are adjured to go to the mountain and return next year as bears; they are then strangled and eaten.

In many parts of the world, in ceremonies involving the ancient custom of ceremonial sacrifice of the king, dogs were used in his place. Among the Huron Indians, dogs were burnt with torches and dedicated to the war god; they were killed and sacrificed instead of the chief. Dogs may also be used in magic ceremonies in which sins and illness are transferred from persons or communities to the dog. In China there was a Taoist (i.e. the doctrines of Lao Tse, c. 500 BC) ceremony for the expulsion of spirits inducing suicide in which a small black dog had his tail cut off with a chopper and was walked bleeding into all corners of the house; he was then kicked howling into the street and it was supposed that the noise drove off the evil spirits. The priest would then burn sulphur and other substances in the house to purify it.

In Hindu countries, dogs are supposed to be reincarnations of men who have committed sins. Egyptian philosophers and even the sophisticated Pythagoras taught that the soul after death went into various lower animals. Pythagoras, as seen above, is said to have held a dog over the mouth of a dying disciple to receive his spirit, because he reckoned that there was no animal so fit to receive the virtues of his deceased disciple as the dog.

The dog has also had a place in rainmaking ceremonies and as a corn spirit. When the wind blows through the corn, the Slavonic peasant used to say 'the mad dog is in the corn', or 'the big dog is there'. In some parts of France they say when a weakly harvester lags behind the reaper 'the white dog has passed near him', or 'the white bitch has bitten him'. In the Vosges district, the harvest may – a large branch or tree decked with corn – was carried with the corn to the house or barn and left until the following year; this branch was called 'dog of the harvest'. The person who ate the last handful of corn 'killed the dog'. Around Verdun when they finished the reaping they used to say that they were going to 'kill the dog'. This may be a relic of the times when the Romans sacrificed a red-coated puppy in the spring to avert the blighting influence of the dog-star Sirius.

No doubt in former times the dog was actually killed and in even earlier times the kings or gods were killed themselves until they cunningly substituted a dog in their place. Sir James Frazer [26] in *The Golden Bough* relates that the dog is a beneficent spirit identified with the spirit of the corn and its death corresponds to the ritual killing of the corn spirit.

In witchcraft and demonology too the dog had a place. As a familiar, the demon dog, of coal black colour, possessed a higher social rating than the cat. 'Wild huntsman' legends are encountered all over northern Europe; the wild huntsman is accompanied by a pack of hounds of fierce demeanour, black and fire-breathing, which are seen by lonely wanderers over the moors at night. These legends abound in western England, especially on Dartmoor, where the hounds are known as 'dogs of hell' or 'whisht hounds'; Windsor Forest is also said to be haunted by Herne the hunter and his pack of hounds. The same idea was present no doubt in the mind of King David when he wrote in Psalm 22.20 'Deliver my soul from the sword; my darling from the power of the dog.'

Chapter 10

Dogs in Hunting and Sport

FROM earliest times until the present, sport and hunting have placed special demands on dog breeders. No doubt at first man relied on an extended development of the herding instinct: hunters and dogs would drive herds of animals towards slingers or bowmen concealed in positions where the animals could be waylaid and attacked or trapped in pits. Even in Palaeolithic times men would drive animals into traps or over precipices; and later the use of dogs in such activities would be a natural extension of traditional hunting methods which, as we have seen, were practised by wolves themselves on occasion. A further development of such activities would be to drive animals or birds into nets. Birds could also be flushed so as to make targets for arrows or fowling pieces; or they were swooped on by hawks or falcons. To this day in Africa, bands of men and dogs frequently drive game, especially wild pig, into nets, where the animals are despatched with spears.

Such ancient methods of hunting are still practised and demand little in the way of specialized properties on the part of the dogs.

On the other hand the ancient Egyptians and Assyrians used dogs to 'course' or run down the prey and in this form of hunting it was the dog which attacked the animal, not the man; this called for special properties of fleetness, aggressiveness and obedience on the part of the dogs, and special breeds would be necessary for the purpose. Further, it would be necessary to breed dogs of different sizes for the hunting of various kinds of animals. The Assyrians, for instance, developed for hunting and war the enormous mastiffs depicted on their monuments.

From ancient Egypt and the desert countries of Arabia came the progenitors of the sight-hunting hounds of greyhound type. Today, as in ancient Egypt and Greece, dogs of greyhound type are the best for coursing and racing; they are also endowed with a useful sense of smell which they use to locate their prey, and afterwards pursue it by sight with such tremendous speed that even the fleetest of wild animals can hardly outpace them. They can be used singly or in packs and, as in the case of the borzoi and other wolfhounds, can be bred to have enough fierceness and courage to tackle the most dangerous of animals, including wolves. As mentioned earlier, these dogs are pre-eminently suited to hunting in desert country or open plains where their powers of sight can be of use, and it is significant that in western countries they have been used mostly in Ireland and Scotland, where there are large areas of open heath; in more wooded country, as in England, their usefulness has been more limited. To enable the landowner to enjoy the thrills of hunting with these animals, 'chases' were cut through the country so as to give the hounds a clear run after the hare or deer. The chase remains as a descriptive name in various parts, though the original use has long since been forgotten.

Among the northern dogs, only one group of large hunting dogs has been developed, those of the elkhound type. The chief use of the northern dogs has been in the realm of animal herding, for which their talents are unchallenged.

The terriers, developed mostly in Britain, were used originally as 'ground dogs' to go underground after fox or badger. In the last century, fox terriers would always run with the hounds and when the fox went to earth, the terrier would be sent underground to tackle it or drive it out. For this purpose they needed to be courageous and pugnacious and many people will remember the rather quarrelsome and untrustworthy fox terriers, very much one-man dogs, of a generation back. Apart from their use underground, some breeds, for instance the sealyham, were used to hunt small animals such as otters. Terriers were also widely kept for the control of rats and mice and a number were usually to be found around farms; they were extremely adept in pouncing on rats and killing them quickly.

The great group of sporting dogs, comprising the scent-hunting hounds – pointers, spaniels, setters and retrievers – were derived from mastiff breeds, as described in chapter 8. They are thus allied to the large dogs of the world, not only the mastiff itself, but also the St Bernard and great dane. By selective breeding and crossing dogs of this type with other groups, it is possible to produce strains with a great diversity of characters. The basic type from which these breeds were derived was probably similar to the pointer-type dogs already described. This basic type was probably a general purpose creature, used among other things for looking after the households and flocks and herds. No doubt they were in part at any rate ancestral to the true spaniels such as the old English water dogge and the

Breton spaniel, and through the spaniel to the setter. Another line of development would be via dogs of Pyrenean mountain dog type, which itself is a kinsman of the ancient large dogs such as the Molossian and Hyrcanian, directly descended from the mastiffs.

The types of dog developed in different countries, in various types of terrain and in different ages, were related largely to the needs of the hunter and the type of hunting. In open country the sight-hunting greyhounds would be supreme because of their superior speed. In countries such as Britain, where in early days the people lived largely on open downland and the wild life was concentrated in the lowland forest and in bogs and marshes, general purpose dogs like the land and water spaniels would be invaluable. These dogs would enter thickets, marshes or rivers to flush game for driving into nets, or for the hawk to pounce upon. Hunting in such countries was with hawk and hound, the hawk being followed either on foot or on horseback. In former times wolves, bears,[1] and wild boar abounded; flocks and homesteads had to be protected and large, fierce dogs of mastiff type were maintained for hunting these animals. Such conditions lasted in England at least until Norman and Plantagenet times.

Originally hunting was pursued for purposes of controlling dangerous or destructive animals and for providing food. It was no doubt also enjoyed as a sport, but the sporting element was secondary to that of necessity. By gradual stages, as the land was cleared and as the more dangerous animals disappeared from the rural scene, hunting became a sport rather than a necessity. Game laws were designed to preserve animals rather than to ensure their destruction for the preservation of property. The Normans brought to England the 'slow-hound', a slow-moving bloodhound, very useful for hunting in thickets by scent. Up to this time, and indeed until the eighteenth century, hunting by packs of hounds was a rarity, since there were no open farmlands. The 'slow-hounds' were taken into the forest, usually with a single leader and followed by the others coupled in pairs; hence to this day huntsmen speak of 'couples of hounds'. The leader was put in first to find the game, then the other hounds would be released, a pair at a time, to follow and drive it. Conditions under which stags were hunted at the time of William Rufus are probably not too unlike those which still exist in the New Forest. One can imagine the huntsmen passing through the forest with their coupled hounds, the hunters on horseback or on foot with their bows and arrows, giving chase behind the hounds until the quarry reached a clearing where it was possible to get a shot.

[1] The bear was abundant in England at least until Roman times (Southern, H.H. [57]) and existed in England until the eighth century AD (Harting, J.E. [28]).

This type of hunting must have been highly hazardous for the hunters, with arrows flying around in all directions; it is not altogether surprising that King William Rufus met his end from an arrow destined for a hart. These hunting methods are strangely similar to those described by Xenophon (see chapter 1).

The hunting of birds with hound and hawk continued through mediaeval times; probably it was something of a status symbol to be seen in the countryside with hound at heel and hawk on wrist, with a hunting horn slung over the shoulder. The dogs involved had by then been differentiated into a number of breeds. The general purpose springer-type spaniel was used in thick coverts; as his name implies, he would 'spring' or flush the game. He was also a good water dog and big enough to bring back a wounded goose from a river or lake. Such dogs could also walk up pheasants or partridges and drive them into nets. They would also serve as satisfactory retrievers, although in this respect they lacked the finesse and soft mouths of the modern retrievers. A smaller spaniel was useful for routing around in thickets and was especially good at putting up woodcock; their use in connection with woodcock earned for them the name of 'cocker' spaniel. The cockers of those days were very similar to their modern descendants.

With the invention of the muzzle-loading fowling piece, a new kind of dog was required – one that would lead his master silently and cautiously to where game lay hidden and indicate its position. Such an animal was already available in the 'pointer', a dog of medium size, much like the prototype from which the sporting dogs were derived. Their senses of hearing, smell and perception generally were highly developed. Their movements were very slow, very precise, very silent; they could creep through the undergrowth causing virtually no disturbance, discover where game lay and lead their master up to it. Having discovered the game, they would stand, with muzzle 'pointing' to where the game was hiding, one foot raised, immobile and rigid. Thus they would stay until their master could creep up, the game would be flushed and the sportsman would have a chance of using the single shot in his fowling piece. It was, of course, important that he should not miss, since it might take him five minutes to reload his blunderbuss once he had discharged it. The whole proceeding was somewhat unsafe, since it was not uncommon for the barrel of the firearm to blow up when it was discharged and the left hand of the hunter could not safely be advanced along it for the purpose of steadying it, for fear that this might happen.

With the improvement in these weapons, a faster-moving and more intelligent dog was required. This was bred from the spaniel in the form of the 'setter', which was much quicker in his movements and allowed for the higher rate of fire which became possible with the development of breech-loading firearms. As in the case of the pointer, the setter would lead his master to where game lay; he would then lie down or 'set', with his body aligned towards the game. When his master arrived, he would at a sign move forward and flush the game to allow his master to get a shot at it.

By the eighteenth and nineteenth centuries, the population in Britain had greatly increased and the land had been largely cleared for farming. Hedges had been planted and the countryside had an appearance similar to that which we know today. The old hunting pursuits of a single man with his dog and gun (without hawk but still with ferret) continued, but new patterns developed also.

Small shoots similar to those of today, in which partridges or pheasants would be 'walked up', were organized by farmers and landowners; the guns would be in line and the birds shot as they flew away. Then came the necessity for trained dogs which could spot where birds fell and retrieve them. Hence the development of the various retrievers, of which perhaps the most popular today are those derived from the dogs of Labrador, both curly coated and smooth, and also now the golden labrador and other colour variations. These dogs could be trained successfully to walk to heel, to spot where a bird or hare fell, to move forward and retrieve the kill only at their master's command, and to bring back the prey undamaged in their soft mouths. They will also work satisfactorily in water, being hardy, immune to cold, and good swimmers. The nineteenth and early twentieth centuries saw the development of the great 'battues' or shoots of driven birds in which each man would attend with his pair of Purdy guns and his loader, taking a right and left as the driven birds approached him and another right and left with his second gun as they receded. Skilful shots would bag four birds within seconds and the barrels of the guns become almost red hot. The role of the dog in this kind of slaughter is not important.

It is not clear when the practice of hunting with packs first became popular. Probably this was a slowly developing process as more and more country became cleared and open. Possibly in the original open areas, as on Dartmoor, in Yorkshire, and in John Peel's county of Cumberland, hunting with packs was developed earlier than elsewhere. This would indeed seem probable from local legends of the black huntsman with his hounds of hell. There may well have been packs of scent-hunting hounds controlled by huntsmen and horn from mediaeval times; and the packs may have been followed either on foot or on horseback.

The earliest hounds involved in such pursuits were beagles and basset hounds. The basset hounds received their name at the end of the sixteenth century, being known previously as 'chiens d'Artois'. They were bred in order to obtain hounds with the hunting instincts and abilities of swifter dogs, but with powers of running restricted to enable sportsmen on foot to follow them. They are very popular for this reason on the continent. Hounds were – and are – controlled by horns; it seems probable that horns were in use in very early times and certainly were the invariable accompaniment of huntsmen from Norman to mediaeval times. Huntsmen used them not only for controlling the hounds but also for communicating with each other. It was the practice in early days to mix packs of hounds in such a way that the huntsmen could recognize the voices of the different dogs. It is characteristic of these hounds to hunt by scent and when in full cry to give tongue, indicating the direction in which they are pursuing the quarry. The huntsmen can then follow the sound if the hounds have disappeared from sight, and can control them with the note of the horn. Such hunting, of course, requires great skill on the part of the huntsman.

Over the years, hounds of different sizes, suitable for different purposes, were developed and the hunts gradually took on the flavour of social occasions. Originally, as today, the local landowners and farmers would meet together and form a hunt, both for their own enjoyment and to control foxes and other vermin when they were becoming too numerous. Fox hunting became more of a social occasion in the late eighteenth and nineteenth centuries and a certain 'status' was associated with a well-known pack. The thrill of hunting in the more favoured counties such as Leicestershire lies in the great speed at which both fox and hounds go away, requiring great nerve and intrepidity on the part of the mounted hunters to keep within distance of the hounds. It involves skill in riding, courage, high-class horsemanship, and great excitement. Today when the ethics of this kind of hunting are called in question by many humanitarians, one may reflect on the paradox by which organized hunting probably preserves rather than destroys the quarry and that without it the fox might become as extinct as the wolf is in Britain.

Various arts of hunting have contributed greatly to the breeding and shaping of our modern dogs. From time immemorial, man has felt the thrill of the hunt; this instinct is built into him, derived from predator ancestors. The excitement of the chase is still an inseparable part of his nature. In most forms of hunting, the dog has been an indispensable asset. He has been bred into many shapes and forms, with divers talents which enable him to take his place in any form of hunting, whether to drive animals for pit, net or hawk, or towards concealed hunters; to course the wild horse, tiger or hare; to hunt the fox with packs of hounds in full cry; or to bring back dead and wounded birds from copse, lake or marsh. Perhaps more than any other influence, hunting in its various forms has dictated the development of canine breeds all over the world and in all periods.

This rapid survey of the changes in hunting customs over the past thousand years or so in England is but a sample of what has taken place here and elsewhere. Much of the material is taken from *The Hunting Instinct* by Michael Brander [5], who describes the changes of hunting methods, mostly in Britain. Surely a whole encyclopaedia could be devoted to hunting fashions of man and dog in the different countries of the world.

So far, we have attempted to survey dogs in the past and present; to form an impression of how and why the different breeds have arisen; and to indicate the relationships of dogs with man. We have seen how man originally adopted ancestral canine species, perhaps with reluctance and certainly only for their usefulness; and how as a result, the dog entered more fully into man's regard and affections and took part in many aspects of his life that seem remote from the original purposes of this association. The final part of this book surveys the natural history of the dog, in connection with his origins from more primitive animals; his relationships with other wild canine species; and his instincts and ways of life as compared with those of his undomesticated cousins. In studying the features which distinguish our modern breeds, we shall see how these are inherited, how the original characters have diverged, and how breeders set about enhancing characteristics they wish to develop.

Part Four

The Zoology of the Dog

Chapter 11

The Relationships of Wild and Domestic Canidae

ALL members of the Family Canidae grouped in the genus *Canis* share many unusual characteristics and have no valid distinguishing features which can justify the creation of separate genera. Such distinguishing features as they have grade one into the other, so that there is a continuous series bridging the gap between each group. The justification for this statement, which can also be applied to all the breeds of domestic dogs, will appear in the succeeding pages.

Canis may, therefore, be regarded as a 'scatter' of phenotypes within one polymorphic genetical system. The members of this genus have a single genetical constitution, within which emphasis on different characters varies in the different groups. This generalization applies with the greatest force to the wolf, jackal, coyote, fox and domestic dog, though it does not mean that certain mutations have not occurred from time to time. Among the wild members of *Canis* these have evidently not persisted, but in the domestic dog there are a number of unfavourable gene mutations present in some breeds which have been preserved artificially by man. Such animals would quickly disappear under the harsh conditions of selection in the wild, and their persistence in the domestic species does not shake the argument.

At some stage it may be supposed that there was a group, or groups, of primitive Canidae which had in their genetical system the capacity to evolve into *Canis* with its very remarkable potentiality, and also into other genera derived from the original. Such variability as *Canis* shows is perfectly possible within any genus where the gene pool is sufficiently extensive.

A similar concept in regard to the genus *Homo* was put forward by Carleton Coon [15] in his book *The Origin of Races*. He considers that early forms of man were not essentially dissimilar from modern *Homo sapiens*, because man's pool of genes is able to produce primitive features such as hairiness, large canine teeth, and heavy supraorbital ridges. He also makes the valid point that a continuous series leads back from modern man through primitive races such as the Australian aborigines to the Neanderthaloid and Pithecanthropoid types. He emphasizes the importance of such a system of neoteny, or throwback to more primitive characters found in earlier types and retained in young animals. Thus some features both in domesticated specimens of *Homo* and *Canis* are in fact those of child or puppy and have been retained by a form of arrested development in those particular features. In man, this applies to the condition of the skin, teeth and skull; in dogs to the condition of floppy ears, short legs and shortness of muzzle.

The concept of predetermination, which is evidently a feature of evolution, is expressed by Edward Fitzgerald in his translation of the *Rubaiyat of Omar Khayyam*, stanza 53, which reads as follows:

> 'With Earth's first Clay They did the last Man's Knead,
> And then of the Last Harvest sow'd the Seed:
> Yea, the first Morning of Creation wrote
> What the Last Dawn of Reckoning shall read.'

However, let us leave philosophy aside and study first the general characters shared by the Canidae which have been derived as variations on the central genetical theme. After this we shall study some of these characters in greater detail.

Dogs, jackals, wolves, and foxes all belong to the Family Canidae, Order Carnivora, Class Mammalia. Other Carnivora include the Ursidae (the bear family), the Mustelidae (stoats, weasels, and otters), the Procyonidae (raccoons), the Ailuridae (pandas), the Viverridae (genets and civets), and the Felidae (cats).

The Canidae have certain rather surprising characteristics, and are distinguished by curious abnormalities of structure common to all of them. In spite of this close structural agreement, they have a great diversity of form. All Canidae resemble either the common wolf or the common fox, though with much divergence in size. In different groups the legs may be longer or shorter; tails may be shorter than the wolf's, though never longer than the fox's brush; the ears are sometimes very large, but are always erect except in some breeds of domestic dog and in puppies; the colouration of wild species varies from grey to yellowish or reddish-brown, though some arctic species are white, and black varieties are not uncommon; the back, upper surface of the head and some parts of the limbs are generally darker than the flanks; the underparts are always paler or even white; the tips and inner parts of the ears may be white, though the external aspects are coloured. As with many other characteristics, the colouration is variable within a single species. The coat may be longer in winter and lighter in colour or even white.

The variability of Canidae is not confined to differences between genera or even species: the members of a single pack or even litter of wild members of this family, particularly those from which the domestic dogs are derived, vary very widely in conformation, colour and temperament. In earlier chapters comment has been made on this great variability in relation to the segregation of breeds and the conditions of domestication. Variability of this kind must inevitably be present in social animals which depend for their way of life on pack organization and a hierarchy system. Complete uniformity would not permit such a system to work among dogs any more than among men. Thus biology gives the lie to the theory that all are born equal.

Apart from this genetical variation, within small limits there is uniformity in all members of the genus *Canis* in respect of dentition and other features of the skeleton. On the basis of anatomical features, only five genera of Canidae can be accepted: *Icticyon* (*Speothos*), the bush dog (one species); *Lycaon*, the Cape hunting dog or hyaena dog (one species); *Otocyon*, the large-eared Cape dog or bat-eared fox (one species); *Cyon* (*Cuon*), the dholes (two species). All others – dogs, wolves, jackals, and foxes – are included in the genus *Canis*. They have no features which could justify the creation of additional genera and they all show many diversities which give an intermediate range between the different species, so that the validity even of speciation can be questioned. The wide distribution of *Canis* is most remarkable and argues for a recent origin from the ancestral form.

In the first chapter, supposed differences between the teeth of wolf, domestic dog, and jackal were discussed. These differences did not appear to support the rather sweeping statements that have been made to the effect that jackals could not be ancestral to domestic dogs because of these features. Our own studies of jackal, wolf, and dog skulls have not revealed these differences in the specimens we have examined; it is evident that such features were merely part of the normal variations and they cannot be used in arguments about ancestry.

Among the Canidae there is a common pattern with a range of behaviour varying according to the habitats colonized by different forms. Variations of habit and colour are undoubtedly the results of response to the dictates of the habitat rather than to a change in the genetic constitution due to speciation. Certain genes are thus suppressed but not eliminated, and emphasis is given to those which determine habits and constitution suitable to the way of life demanded by the habitat. Under changed conditions, by a different emphasis new forms and ways of life could be produced within a few generations. It is this capacity, arising from the wide gene pool, which has enabled man to produce so many breeds of domestic dog from the basic material. Various species, such as the wolves, hunt their prey in packs and can be very dangerous both to wild animals and to man; other species, although closely related, have forgone this type of life. The South American wolf, for instance, is not a dangerous animal; it lives in solitary state and attacks only small game. However, the red wolf of Texas (*Canis niger seu rufus*), which is apparently very similar, will combine into packs particularly in winter and is a source of danger to herds of cattle. Other Canidae, such as jackals, live mainly on carrion and young birds and eggs; they will also eat lizards, mice, snails, and insects, white ants, and moths. Other species of wild dogs living along rivers or by the sea will eat crustaceans and molluscs, and arctic foxes and timber wolves eat fish.

Some species of Canidae eat vegetable foods and fruit. It is well known in Britain that the common fox eats an enormously varied diet. At one time it relied largely on rabbits; when these were no longer available because of the outbreaks of myxomatosis, the numbers of foxes did not diminish; they found alternative forms of food, although admittedly they made more frequent raids on poultry and turkey farms. Apart from this they found sustenance by eating invertebrates, roots, carrion and anything else they could find. Thus there is an infinite gradation between Canidae which are almost entirely carnivorous and forms which eat almost any other kind of food, or even some which are herbivorous. In this respect, as in so many others, *Canis* greatly resembles *Homo*. All forms can subsist on carnivorous, omnivorous or even vegetarian diets.

The Canidae pursue their prey largely by scent and their olfactory organs are extremely sensitive and well developed. Their senses of sight and hearing are also acute, though there are gradations between those which rely mostly on sight and those which rely on scent. Most Canidae are active during at least part of the night, though many are abroad also during the day. In hunting, water is no barrier to them; many domestic breeds also take to water and are naturally good swimmers.

An essential part of the social organization of Canidae lies in the sounds they make. In the wild species, these are howls, yelps or growls; only domesticated dogs habitually bark, although dingos, wolves and jackals will learn to bark from domestic dogs if kept in captivity. It is recorded that both timber wolves and pale-footed Asian wolves do sometimes bark in the wild; dingos certainly do, usually at the end of a howl. The habitual wolf howl is the pack signal to draw them together for some combined effort.

Pack hunting involves more than a simple running down of prey in unison; the hunt is organized with some members rounding up the prey, separating an individual from the herd so that it can be attacked by the pack as a whole. In a wolf pack, the dominant wolf is the leader elected by merit and often after fighting; he is not, however, necessarily the strongest and it must be supposed that often he obtains his position by cunning in overcoming a stronger opponent. It is these qualities of organization that have made the domestic dog derived from these wolves such a valuable ally to man; they have also made dogs acquiescent to domestication and obedient to the dictates and instructions of their master, who is probably regarded as a super pack leader. Outside the genus *Canis*, domestication of Canidae may be difficult. Most Indian dholes, for instance, are said to be untameable. Within the genus *Canis*, however, almost all members can be tamed and reared in captivity, and they will interbreed readily with each other. These remarks cover wolves, dogs, jackals and foxes.

It is often said that foxes are distinguished from dogs by the characteristic odour arising from their anal scent glands. This statement is incorrect, since many species of wild dog smell very strongly; and although the common fox has a distinctly offensive odour, the arctic fox does not.

Wild Canidae have a very wide geographical distribution and are present in almost all parts of the world. In the Old World they are distributed from Spitzbergen and Siberia to the Cape of Good Hope and Java (Indonesia). In the New World they are to be found from the Arctic Ocean to Tierra del Fuego and the Falkland Islands. There are twenty species in the northern hemisphere and twelve in the southern hemisphere; three of these are common to both. The species common to both the Old and the New World are *Canis lupus* (the northern wolf), *Canis vulpes* (the common fox), and *Canis lagopus* (the arctic or white fox). Apart from the dingo, which is undoubtedly an introduced form, there are no indigenous Canidae in Australia, New Guinea, Tasmania, New Zealand, Celebes, the Philippines, Ceylon, Madagascar, or the West Indies.

Both wild Canidae and the domestic dog can thus endure great extremes of climate. Man, of course, can do the same; and this, together with his great variability, his adaptability with regard to diet, and his capacity for social organization, is undoubtedly responsible for his wide distribution and success in the biological competition for survival. The same may be said of the dog family.

All Canidae whose habits are known make greater or lesser use of burrows. Sometimes these are simple, but sometimes they are so extensive as to constitute an underground canine village. The extent to which burrows are made for living in or merely for giving birth and raising the puppies depends largely on habitat. Foxes living in moist forested countries and which are frequently hunted and persecuted make extensive burrows for warmth and security, and also because the soil is suitable for this kind of activity. In desert countries and where burrowing may be difficult, jackals, coyotes and desert foxes will find themselves secure resting places among boulders, where they can lie up and bear their young. All genera find some kind of burrow or nest in which to give birth; litters consist of from three to twelve puppies, which are born helpless and blind after a gestation period of 62 to 68 days.

Thus, in spite of their apparent differences, all Canidae are remarkably uniform in their way of life and organization. Such uniformity extends also to their anatomy. In the external anatomy all members of the group are similar to the common wolf, except as regards size and relative length of tail, ears and muzzle. Only in *Icticyon* (*Speothos*) *venaticus* (the 'bush dog') is the tail really short, and only in *Canis zerda* and *Otocyon megalotis* (the fennec foxes and the bat-eared foxes) are the ears excessively long; even in these species the ears do not droop as in some domestic dogs. In wild Canidae, the length and quality of the fur is variable. There is no hallux (big toe) on the hind feet, although a residual fifth toe is often present in domestic dogs in the form of dew claws. These well-formed and sharp claws are vestigial structures usually not connected with the skeletal system by bones or ligaments. As in the case of the drooping ears, they probably reappeared as a result of neoteny; they are usually surgically removed from new-born pups. In true dogs (*Canis*), a short pollex or thumb is present in the front feet though it does not reach the ground. It does not appear in *Lycaon*, but is concealed beneath the skin. In no Canidae are the claws retractile and all existing genera are digitigrade (though one extinct form appears to be plantigrade); this means that modern Canidae walk on the fingers and toes and not on the palms or soles.

All Canidae have seven cervical vertebrae – as in almost all mammals – 13 thoracic, 7 lumbar (rarely 6), 3/4 sacral, and 11/22 coccygeal or tail vertebrae. The clavicle is represented by a small cartilage in the flesh, except in *Lycaon* in which it may be considerably larger.

In the skull, also, remarkable uniformity is shown in all these animals. The zygomata project strongly outwards. An elevated ridge of bone, the lamboidal ridge, crosses transversely the hinder part of the cranium. The bony orbits never make a complete ring, and in wild specimens the nasal bones are always elongated, though this feature is very variable even among the same species such as *Canis lupus*. A sagittal crest may or may not be present, projecting upwards from the middle of the cranium. This crest is well-marked in the wolf, is present in many breeds of domestic dog, and is sometimes present and sometimes absent in jackals; where absent it is replaced by a flattened tract of bone. The existence or absence of this crest may depend on the type of diet which animals receive. It must be remembered that the shape and bulk of all bones depend greatly on an animal's mode of life; in young creatures, bones are plastic structures which can vary greatly according to the uses given to the muscles attached to them. A dog which in early life receives a hard diet will develop powerful muscles of mastication, and these in turn will induce the development of powerful skull bones in those areas against which they work.

Domestication with no out-breeding can produce marked differences in the relative proportions and shapes of the bones of the skeleton within one or two generations, simply as a result of a changed way of life. It is on this basis that archaeologists claim to be able to distinguish between domestic and wild Canidae found on ancient sites.

The tooth formula of dogs is, in each side of each jaw: incisors 3; canines 1; premolars 4; molars 2 above and 3 below. This is expressed:

$$ I\frac{3}{3} \quad C\frac{1}{1} \quad P\frac{4}{4} \quad M\frac{2}{3} \quad = \frac{10}{11} $$

One true molar, that is a tooth without a milk tooth predecessor, is present in all genera in the upper jaw, and there are at least two in the lower. *Icticyon* has no more molars than this; in *Cyon* there are two true molars in both jaws; in *Otocyon* there are three or even four true molars above and four below. In all other Canidae – the vast majority – there are two true molars above and three below. Among all groups abnormalities and exceptions occur, and in some of the more specialized breeds of domestic dogs some rather surprising tooth formulae are found.

In the upper jaw the fourth premolar, and in the lower jaw the first molar, constitute carnassial teeth. In *Otocyon*, however, these teeth do not differ markedly from other adjacent teeth. In the northern wolf, if not also in the Asian wolf, the length of the carnassial teeth exceeds the combined length of the two succeeding teeth. This is said to be a distinguishing character between wolf and dog and wolf and jackal, though this is surely a somewhat invalid distinction.

Compared with the Felidae, the incisors of the Canidae are relatively large in both jaws. In the upper jaw, the first premolar has a single fang; all teeth behind it have two fangs, while the last three upper teeth have three each. In the lower jaw, the first premolar and the third molar each have one fang, while all the intermediate teeth have two fangs.

In a later chapter we shall study in greater detail some of the characteristics of the skull and teeth of wild and domestic Canidae in an attempt to discover features typical of the various groups of domestic dogs. In this we shall be unsuccessful. We shall find, as everywhere with Canidae, that there are a series of variables shading from one group into another. Many anatomists have exhaustively measured skulls and other bones of different breeds of wild and domestic dogs, but no criteria have ever been established by which skeletal material can be differentiated.

A systematic account of the better known genera and species of Canidae is given in appendix I. The classification of the living Canidae is extremely simple:

$$\text{Digits 5-4} \begin{cases} M\frac{1}{2} & Icticyon \\[1ex] M\frac{2}{3} & Canis \\[1ex] M\frac{2}{2} & Cyon \\[1ex] M\frac{3}{4} & Otocyon \end{cases}$$

Digits 4-4   *Lycaon*

As with the rest of the data relevant to Canidae, that from palaeontological sources is confusing. Remains of a supposed dingo have been recovered from Pleistocene deposits. Other Canidae have also been found in caverns, but all in strata dating from the Quaternary period. It is said that relics of the common fox have been found in Upper Pliocene deposits, but it must surely be difficult to state with any degree of certainty that this animal was a common fox and not some relatively undifferentiated early canine specimen. Remains of *Lycaon* are stated to have been found in Upper Pliocene deposits in Glamorgan; and *Icticyon* is said to have been present in Brazil at this same period. *Cyon* has been recovered from Pleistocene deposits in Europe.

This scanty evidence might suggest that the different genera and species of Canidae segregated in Middle or Upper Pliocene times, perhaps when man himself was in course of active differentiation. As regards possible ancestral forms, one *Cynodictis* has been found in the Lower Miocene and Upper Eocene of Europe; this form is more or less intermediate between dogs and civets (Canidae and Viverridae). Another genus, *Amphicyon*, from these deposits had a dentition resembling that of dogs, but differed in the structure of the feet, being plantigrade and in this respect resembling the bears.

Chapter 12

The Ancestry and Characters of the Canidae

THE accounts of the different wild members of the Order Canidae given in appendix 1 are derived mostly from Mivart's [44] *Canidae* published in 1890. Although written so long ago, Mivart's descriptions of the Canidae are still the most reliable and contain a fund of information of every kind. We have followed Mivart in his classification of dog, wolf, jackal and fox in the genus *Canis*, since obviously there are no anatomical or other valid features which could justify placing these forms in different genera. Thus we have not accepted *Vulpes* as a separate genus for the fox. In this we follow also Winge [66] who in his *Relationships of the Mammalian Genera* volume 2 (1941) goes even further, making the Canini a sub-group of Ursidae (the bear family). On the basis of the known ancestral forms, he considers that the bears and dogs have branched from a single root. Winge's classification of Ursidae is as follows:

Ursidae

   I Premolars well developed, of usual form.

     Canini

      (a) First toe well developed.

         Cynodontes: Cynodictis, 'Galecynus', Cynodon, Cephalogale, Daphoenus, Temnocyon

      (b) First toe atrophied.

         Canes: *Canis, Otocyon, Lycaon, Hyaenognathus, Ictidocyon, Eydridocyon.*

  II Premolars atrophied, or were once partly atrophied.

     Ursini

         Amphicyon, Simocyon, Dinocyon, Hemicyon, Hyaenarctus, Aeluropus, Ursus, Melursus.

The genus *Canis* includes *Vulpes, Cyon, Chrysocyon, Palaeocyon* and other ancestral forms. *Otocyon, Lycaon, Ictidocyon, Speothos (Icticyon)*, the 'bush dogs', are given separate generic status, as with Mivart.

Thus bears and dogs have originated from a common form which had five digits on fore and hind feet and walked, in the same way as bears, on the soles or palms (plantigrade); the dogs, as we shall see, lost the thumb or big toe (pollex and hallux) secondarily and came to walk and run on the digits (digitigrade) as a secondary adaptation to rapid movement.

The Ursidae can be traced back to extinct primitive carnivorous animals known as Amphictidides. The lowest Ursids advanced above these animals, but are very similar except for the development of larger tympanic bones, which are saucer-shaped and constitute the entire external wall of the tympanic cavity. Such a development would suggest an improved sense of hearing. An identical character occurs in a number of other families which are also supposed to have been derived from the Ursids, namely the Procyonidae (raccoons), the Mustelidae (stoat, weasel, skunk, otter, etc.), and the Otariidae and Phocidae (sea lions and seals). In the typical Amphictidides, there is no lower third premolar, although this is present in the upper jaw. These teeth, however, are all present in the primitive Ursidae.

In Tertiary times there lived in Europe a primitive form of Ursid known as *Cynodictis*. In this form the teeth were very like those of *Amphictis* and a small lower third premolar was retained; these two forms were also alike in most other features of the skull. The body resembled the most primitive forms of the Viverridae (civets and genets), with rather short limbs, five fingers, and five toes. An allied form lived in North America and two others closely allied to *Cynodictis, Cynodon,* and *Cephalogale*, occur in Europe. In these the clean-cut shape of the molars, which was a feature of primitive Carnivora, tended to be effaced, as it is in our modern dogs. Indeed in *Cephalogale* the first lower molar, which it will be recalled is the lower carnassial of Canidae, has begun to show modifications. As regards dentition, in Winge's view *Canis, Cyon, Chrysocyon, Palaeocyon*, etc. could not be generically differentiated from these early forms. The skulls, too, show resemblances to *Cynodictis, Cynodon,* and *Cephalogale;* but they are readily differentiated by reason of body shape, since the early forms were squat and evidently plantigrade.

Canis, with its long limbs, is much better adapted for running and is pronouncedly digitigrade. From *Canis*, the few genera distinguished have evolved in different directions, although all the forms of *Canis* proper are closely related and have probably separated only in recent times. Rather generalized forms of Canidae (Canini) probably existed over long periods and were widely distributed in all parts of the world. Segregation into more specialized forms has taken place at various times and in various places, but without the formation of new genera. The differences between modern and extinct forms are extremely slight and are related to a greater or lesser specialization as carnivore or omnivore, as sedentary or active running animals, and in relation to the development of various senses – hearing, scent, and other differences associated with habitat and way of life.

Thus there existed towards the end of the Miocene period a small, unspecialized, carnivorous animal which moved rather slowly, walking on palms and soles; it lived and reared its young in burrows; and it was able to revert to other forms of sustenance if prey was difficult to obtain. This little animal was apparently rather defenceless and its young were born in a very helpless condition, as are those of this group of Carnivora today. Probably many succumbed to the hazards of life and so they were born in multiple litters. Evidently they evolved to fill some niche, but they had little future unless they became further specialized under the influence of natural selection to provide more efficient types. These creatures lived some twelve million years ago, when the Miocene period was verging into the Pliocene.

From them emerged various forms better equipped to survive in the harsh conditions of natural competition. One group, the bears, became large, blundering animals, well able to protect themselves. Another group became very small, short-legged and savage; these were the Mustelidae, comprising stoats, polecats, weasels, wolverines, skunks, badgers, and other such animals, including one semi-aquatic group, the otters. Some members of the group evidently had a liking for water and from them were derived the seals, sea lions, and walruses. From this stock, too, came the Ailuridae or pandas; also the Viverridae, including the civets, palm civets, genets, mongoose, and meerkats. Perhaps the most unspecialized of these groups and the one most closely resembling the ancestral form is the raccoon family, the Procyonidae, consisting of the cacomistles, coatis, kinkajous, and the raccoons proper.

Each of these groups radiated into different ecological conditions, giving them a specialized habitat in which they could become the most important predatory animal and thus ensure survival. Alone among the descendants of the primitive *Amphictis* group of Tertiary mammals, the Canidae developed predatory habits requiring strength, speed, ferocity and other properties such as good sight, hearing and scent; by virtue of these qualities they could detect their prey and hunt it down by superior speed. These properties do not exist in any other animals except perhaps the cheetahs, which also possess both speed and stamina. The other main group of Carnivora, the Felidae, hunt their prey by stealth and rely on great speed over a short distance to catch it; they cannot sustain a prolonged hunt as can the predatory Canidae.

Together with these physical characters, the Canidae possessed instinctive behaviour patterns which were also advantageous to the way of life which they adopted. These are the qualities of organization, both in connection with combining in packs to hunt their prey, and also in the highly developed social and territorial behavioural patterns which persist so obstinately among domestic breeds of *Canis familiaris*. A further characteristic which has gone hand in hand with these developments is the lack of uniformity which is so obvious even in a single canine family. Even in inbred lines, there are puppies of different shapes, colours and other physical attributes, and also with markedly different characters. This property – let us repeat it – has made possible the social organization which is so evident in wild Canidae with their pack leaders; among working dogs in which selection of the team leader for the sledge is so important; and among even the most closely inbred domestic dogs, who always seems able to select the dominant and the subordinate individuals.

As we shall see in the next chapter, the Canidae have the largest number of chromosomes of any known mammalian species, seventy-eight in all. Whether the great polymorphic genetical variability of these animals is connected with the large number of chromosomes they possess can only be a matter of conjecture. It is certain, however, that this great variability, adaptability and ability to form social links with man has endowed dogs with the useful qualities which have made them so valuable to him.

These same powers of variability have also led to the segregation of Canidae into the main groups: *Canis* and related genera on the one hand; and the different species of *Canis* itself – the wolves, jackals, foxes and domestic dogs – on the other. It must be emphasized again that these animals have no basic differences, either in physical or mental attributes.

Whatever the common ancestor was like, the development of great physical powers of speed and endurance, associated with anatomical and psychological specialization, must have been a response to the urge to combine in hunting down prey at speed. The adoption of more sedentary means of acquiring a livelihood by scavenging or gleaning, as in the case of the foxes, must have been secondary, and thus the ancestor was surely more wolf- or dog-like. However, all these groups shade imperceptibly one into the other and, as we have seen with wolves and jackals, many of them exchange one way of life for another and all can exchange their wild way of life for domestication. The varying attributes seen in the various groups of Canidae are therefore all part of a single genetical spectrum, of which one part is emphasized in one group, another part in another group; it can only be supposed that wolf-like animals could be bred in time from foxes and fox-like animals from wolves.

It seems, therefore, that the modern Canidae have been derived from wolf-like ancestors and, on the basis of geographical distribution, the northern sub-arctic regions of Europe may be suggested as the location. If this is so, it is not difficult to account for the distribution of canid forms elsewhere. The passage to the Americas by the land bridge at the Behring Straits during the glacial period would be simple and the distribution of other canid species throughout Europe, Asia, Australia, and Africa is easily explained.

The psychological and anatomical features of the various canid species were already present before man began to select wolves with special characters to serve his own purposes. By selective breeding, man has intensified certain characteristics; but he has not created anything that was not there in the ancestral animals. As small predators in a world of large prey species, the canine ancestors had two alternatives: to confine their hunting to small, easily caught animals and invertebrates (as do the foxes) and to scavenging (as do the jackals); or they had to combine in packs to hunt their prey. It was this necessity which led canine animals to organize themselves in social groups and to develop their advanced social and territorial system, as well as their peculiar codes of etiquette when meeting either an old canine friend or a stranger. Much has been written about the social and other habits of dogs; outstanding among such works are those of Konrad Lorenz [40] and R.H.Smythe [55]. It is not our purpose to write further on this subject, except in so far as it has a bearing on canine natural history.

In our studies of the characteristics of different groups,[1] we have been unable to find hard and fast characters which could differentiate between dog, wolf, and jackal; indeed we have been unable even to confirm the differences of tooth structure in these animals which had apparently been accepted up till now (but see reference to Pocock, p. 14). It seems to us that the answer to the problem lies in the diversity to be found among individual breeds and species rather than in their uniformity.

[1] Some of our results are reproduced in a series of silhouettes and photographs showing the teeth and other features of skull formation in plates 36–40.

The first two silhouettes in plate 36 are taken from the skulls of wolves, both males, from the Zoological Society's collection at Whipsnade. The differences between the two are so striking as to make it clear that there is considerable dimorphism even among wild wolves. One skull clearly tends towards an alsatian/collie type of animal, with long, straight muzzle and limited depth of both mandible and cranium; the other skull tends towards the mastiff type, the skull being thick and coarse. In the first skull the incisor teeth meet cleanly and exactly; in the second they do not meet uniformly, those of the lower jaw slightly overlapping those of the upper. Thus among captive northern wolves we find specimens resembling two of the main groups of dogs.

Let us therefore consider the possibility that among wild wolf packs, both sheepdog and mastiff-type animals are already present; and let us suppose that this dimorphism is related to the social habits of these animals in hunting. The lean-skull type of wolf, probably endowed with greater intelligence, would provide the leaders and those animals whose function would be to separate the animals selected as prey; they would be endowed with the properties suitable for sheep dogs and early man would have had no difficulty in segregating these types by artificial selection. The other type of wolf would make up the shock troops of the pack; these would be endowed with good scent for following herds of prey animals in thick country, and with powerful teeth and jaws to overcome very large animals.

The effects of domestication are seen within one or two generations in the progeny of wild wolves brought into captivity. They arise from two different origins: one due to a drastically changed way of life; the other resulting from the preservation of the smaller and weaker litter mates which would not survive under wild conditions.

In a growing animal, the conformation of the bones of the skeleton is to a certain extent dictated by the degree of use to which the muscles attached to the bones are put. Therefore animals which are habitually given very hard food develop powerful masticatory muscles and powerful bony ridges, such as the sagittal crest, against which the muscles pull. Similarly, animals accustomed to travelling long distances, hunting large animals, and other forms of severe physical exertion (as might be undergone also by sheep- and shepherd dogs) develop strong limb muscles and limb bones and a robust frame generally. Animals reared in captivity tend to lead sheltered lives and thus the skeleton may be less well developed and lack prominences such as the sagittal crest; these differences, which arise merely from differences in environment, may be wrongly attributed to genetical changes, although the basic constitution of the stock is unchanged.

The survival of the smaller litter mates and their use in breeding inevitably leads to the survival of weaker and smaller stock; this in time must have an influence on the race, which will become physically weaker. In addition, the stock will tend to become less sagacious, except in so far as it is bred for special tasks involving some form of intelligence. Changes may also be expected in relation to the teeth, because under conditions of domestication dogs are usually given soft or cooked food to eat. The result is that the teeth are not given the full use for which they have been evolved. They become coated with tartar, the animal suffers gum decay, and the teeth themselves deteriorate progressively throughout life. Teeth, being weapons of offence and defence, are necessary to survival in the wild; but under conditions of domestication, animals will survive which have a poor dental equipment. Under natural conditions, they would probably have been eliminated before they were able to breed. It is not unusual, therefore, to find domesticated dogs equipped with relatively weak teeth and jaws, except when they are bred especially for purposes of war or for fighting other dogs or other animals.

A study of the plates showing the skulls and teeth of wild and domestic Canidae makes abundantly clear the truth of what has been said above. The skulls of the northern wolves are massive; the temporal regions and sagittal crest where the main muscles of mastication are inserted are very strongly developed; the jaws are heavy and strong; the teeth are sharp, in excellent condition, massive and deadly, and the incisors are well developed. A feature not described in textbooks on the anatomy of the Canidae is the tendency for the corner upper incisor to be prolonged, as if it were a third canine tooth. This long tooth can be seen to work against the canine of the lower jaw, which on its posterior side works against the canine of the upper jaw. Particularly in the wild wolves and more primitive domestic dogs, this trio of sharp, pointed and deadly weapons on either side of the jaw forms a formidable assemblage resembling the steel teeth of a hunter's trap. An examination of the plates will show the steady deterioration of this mechanism in the more advanced groups of dogs, which have lost the necessity to gain their living by hunting. The primitive, wide-spaced tooth formation, together with large, sharp carnassials, is well illustrated in all the wolves, coyotes, jackals and also in the dingos and pariah dogs. In foxes the arrangement is less powerfully developed, perhaps because of a life of scrounging and attacking only small creatures.

In these respects, therefore, the anatomy of the skull emphasizes the essential oneness of wolves, coyotes, jackals and wild and domestic dogs. It supports the view that the more primitive dogs, such as the dingos,

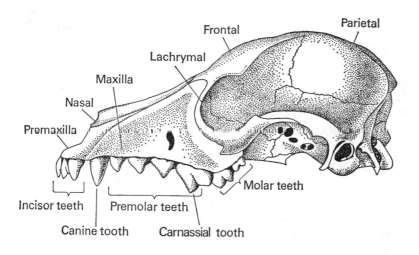

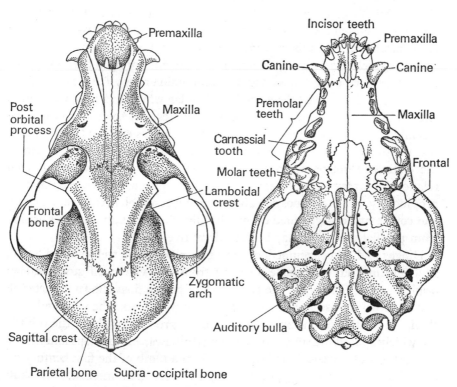

Figure 11. Skull of domestic dog (*Canis familiaris*). (After Collinge.)

huskies, basenjis, and pariahs are very closely related to wild canid ancestors and indeed cannot in any scientific sense be separated from them. It also supports the supposition that the jackals are merely wolves, which have taken to a different way of life and have thus come to develop qualities suited to it within the original genetical constitution.

Various workers have made much of the presence or absence of a 'stop' in the different canid groups. It is sometimes stated that this feature is related to a dog's sense of smell, because within the swelling of the frontal bones which comprise the 'stop' lie the frontal sinuses. These sinuses trap expired air on its way to the exterior, and it is argued that they are therefore associated with the sense of smell. This reasoning is fallacious, both because an organ of smell would require to trap inhaled and not exhaled air, and because the lining membranes of the sinuses do not contain the type of nerve endings which are associated with a sense of smell. The explanation is plainly quite different: these sinuses contain exhaled *warmed* air and thus form a warm cushion over the delicate tissues of the eyes and forebrain, helping also to warm the cold inhaled air as it passes along the nasal passages. For this reason, the 'stop' is best developed in those Canidae which come from the cold northern regions or from the mountains, such as the northern group of dogs and mountain groups like the mastiffs.

The 'stop' is less well developed in Canidae which live in warmer areas and is absent in greyhounds, jackals and coyotes. These groups bear some resemblance to each other, and also to other species such as *Canis simensis* and the dholes (*Cyon*). These resemblances, which can be clearly seen in the plates, do not necessarily denote a close relationship, although such relationships may nevertheless exist in those groups where geographical propinquity makes this possible. The striking resemblances between jackal and coyote, widely separated geographically, show that appearances arising from parallel evolution may be very deceptive.

Dingos and other members of their group possess a well-marked 'stop', probably indicating a single ancestry from one wild wolf stock and an origin in colder climates. The two skulls of pariah dogs shown in plates 38 and 40 reveal the essential characters of wild wolf stock; the wide and deep skull, the width over the zygomata (the bony projections above the orbits), the widely spaced incisor teeth, the powerful jaws, and perhaps a tendency to shortness of muzzle. The little skull from a tomb of the first century BC at Denderah in Upper Egypt (plate 40) is of very great interest. This skull was presented to the British Museum (Natural History) by the great egyptologist Sir W.M.Flinders Petrie and was ascribed either by him or by museum officials to the group of pariah dogs. Nevertheless, it had been buried with ceremony in a prepared grave and, although exhibiting characteristics resembling semi-wild creatures, it was more likely to have been a household dog of basenji type. The skull was of a dog, almost certainly male, small but powerfully built, which had reached advanced years. From the tooth wear, it had evidently been fed on hard rations. Two of the incisors had been lost during life, but the other teeth, though worn, were in general very healthy. The loss of the incisors may have occurred as a result of a kick or a blow from the hoof of an animal, and possibly this dog was one from a pack used for hunting.

From this evidence and from that of the Indian pariah skull we may assume that in spite of their scavenging habits, these dogs have the equipment for hunting and defence possessed by wild Canidae and that this is much more developed than is commonly seen in domesticated species. One might suppose that the curs which infest Middle and Far Eastern cities would be degenerate creatures; from the evidence of these skulls, this is not the case.

The greyhound group can be clearly differentiated among the plates. The lurcher skull (plate 37E) is evidently of greyhound type; the Nepalese mastiff also shows greyhound features, from which one must suppose that the mastiffs in this area have been crossed with greyhound blood; indeed we have already commented on the early introduction of greyhound blood to northern India.

The short-muzzled dogs are represented in plates 37 and 38 by boxers and the pekingese. In the pekingese, all the lower incisors are missing as a result of non-development. All the teeth are degenerate and set in bones which lack proper sockets to give them firm support. These skull formations have clearly arisen from intensification by inbreeding of characters which are already present among the wild northern wolves. It appears that the length of muzzle and jawbone is separately determined, so that inequalities may result.

Among the features of the skulls especially worthy of note are the shapes of the mandible. In the northern dogs these tend to be rounded, whereas in greyhounds they are rather flat, as also in the jackals. It seems almost as if the mandibles adjust their length to the maxillae by producing greater or lesser curvature; thus in the long-muzzled dogs the mandibles have flat lower edges, whereas in the short-nosed dogs the lower edges are rounded like the keel of a ship.

Finally, the skulls differ in relation to the development of the auditory bullae, those rounded projections below the ear which contain the tympanic cavity. These appear to be better developed in the jackals and dogs of the greyhound group than in the northern dogs. It would therefore seem that in the sight-hunting groups the sense of hearing is better developed than among the scent-hunting groups. Expressed in a different way, those groups which hunt in thick country appear to rely less on sight and hearing and more on scent; those which hunt in open country appear to rely more on sight and hearing. These faculties are revealed in the anatomy of their skulls. Not only are the auditory organs indicative of these habits; in addition, the nasal passages are much wider in the scent-hunting groups than in the sight-hunting groups.

Domestication has had further influences – on the breeding cycle, on developmental characteristics, on posture, and on general habits. Wild Canidae come in season and breed once a year; so do the dingos, pariahs and basenjis. Domesticated dogs come in season twice a year, and twice a year they regularly breed, or attempt to breed. This provides further evidence of the closeness of members of the dingo group to their wild ancestors. However, whereas wild wolves and dingos have no dew claws on their hind legs, these are well-developed in the basenjis.

Whereas wild wolves, dingos and many domestic breeds carry their tails hanging low over the anus, the basenji together with some other domestic breeds have small curly tails, or tails which they carry high. This feature, together with the puppy-like appearance of the ears of many domestic breeds, is ascribed to neoteny. As noted earlier, the tail in wild Canidae may be used to mask or unmask the scent arising from the anal scent glands, a property which might have importance both for survival and for tracing a mate or other members of the group. These properties would have less value under domesticated conditions, and may have been lost in the natural course of breeding. Nevertheless, both the tail formation and the existence of the dew claws does suggest that the basenji at some time, possibly in early Egyptian times, was selectively bred, even though it still retains many primitive features.

Although many explanations of the development of the habit of barking in domesticated dogs have been put forward, it is difficult to explain this unless man has himself bred dogs for their ability to give tongue in connection with guard and watch duties. Wild wolves howl and only rarely emit a kind of bark; dingos regularly produce a bark at the end of a spell of howling. All dogs can learn to bark from domestic dogs, except possibly the basenji. Another problem concerns the means by which silent hunting dogs, as opposed to those which hunt in baying packs, have been developed. Whether this was an atavistic trait among the ancestors from which they were derived, or whether man has developed this feature by intensive breeding, must remain one of the many mysteries associated with the origin of *Canis familiaris*.

If the skull of a wolf, dingo, jackal, coyote, fox, or other wild canid is placed on a table or other flat surface with the mandible articulated, it will rock from back to front. The crossed wolf/husky skull also rocks and so does the ancient Egyptian skull from Denderah. Among the truly domesticated dogs, only the skull of the fox terrier rocks in this way. All others rest leaning backwards on the mandibular process or the occipital bone. The fox terrier is plainly the exception that proves the rule.

The little wolf-like skull from Denderah rouses the mind and the imagination. The superb and nearly intact temple of Denderah was built by Cleopatra for Marc Antony; the walls are adorned with pictures of Cleopatra and her lover and on the roof is a lovely stone boudoir, where the queen took her ease. It is not unlikely that this wild wolf-like animal may have crouched at the feet of this great, wayward, savage woman.

Additional Note

The reference at the top of page 131 to Winge's classification of *Cyon* as a member of the genus *Canis* is not followed by the authors, who adopt Mivart's classification (see page 128).

Chapter 13

Behaviour Patterns

IT has been our contention throughout this book that, with the possible exception of three genera, all Canidae belong to a single very variable group of Mammalia. Even though there may be slight differences of chromosome numbers and morphology in the more extreme groups, in the main groups at any rate all can interbreed and produce fertile offspring. Furthermore, anatomical diversities between individuals suggest a polymorphic geneticism within which all forms of wild and domestic Canidae could be divergent members. The very fact that man, by methods of selective breeding, has produced forms apparently more dissimilar than those which exist in wild Canidae strongly supports the argument that these are nature's own segregated varieties produced in response to the demands of widely differing habitats and ways of life. It was indeed due to the fact that the canine family has the power to segregate in response to natural demand that man was able to produce the enormous variety of breeds which make up the animal known as the domestic dog.

Anatomical variations of structures are not only closely related to ways of life, but also, it is suggested, to specializations of function within the social order or system, including different functions in pack hunting. Thus it seems that in a pack there may be some individuals which have the special capacity to herd and round up animals for the kill, and others of more massive build who in the main do the attacking. We see the projection of these two types exemplified in our domestic dogs, especially in those of collie type which are pre-eminent as sheep- and shepherd dogs, and in those of mastiff type – the massive dogs – which attack the larger animals in the hunt.

Along with this anatomical diversity, can be traced also differences of temperament which are just as important and equally related to ways of life in the wild. Possibly different behaviour patterns are even more striking than anatomical variations. In this respect the work of the archaeologist in determining the nature of a canine specimen represented only by a skull or teeth – or even by a full skeleton – becomes extremely difficult. Nevertheless, studies of fossils do give some indication of the mode of life of the specimen. A long, lean body with long legs may indicate a dog of stamina and speed which was used in hunting. Expanded nasal bones may indicate a hound of the scent-hunting type. Large auditory bullae signify good powers of hearing. The degree of wear and condition of the teeth may suggest the nature of the diet. For instance, well-formed teeth without great wear or signs of disease indicate a well-balanced diet, probably containing a reasonable proportion of meat but without an excess of hard materials.

We have examined such a dog in the Windmill Hill skeleton at Avebury Museum (*vide* appendix III) and may use this to provide an example. After death this dog was flung into a ditch without proper burial and it is therefore unlikely to have been a favourite pet of the family. Nevertheless it was evidently well fed and cared for; the skeleton shows no injuries which could be attributed to sticks or stones or kicks, although some time before death it seems to have suffered from a blow which caused separation of the two maxillae. Although the dog was not a special family pet, evidently it was highly regarded. It had good powers of scent, speed and probably endurance. We know that the owners kept numerous cattle, sheep and goats and we may suppose that this was primarily a sheep- or herd dog, possibly used also for hunting.

In the study of animal groups, much confusion is caused by the habits of taxonomists in creating genera and species on the basis of quite minor morphological differences. Where there is a wide gene pool in which heterozygous elements predominate, there must be great variability and certain characters will be intensified in relation to different ways of life and kinds of habitat. This is well known and needs no elaboration; it arises from factors of natural selection, which emphasize beneficial characters at the expense of those which are neutral or detrimental. The system allows the form to become adapted quickly to new conditions without the creation of new genera or species with entirely new characters produced by the lengthy process of mutation and fundamental radical change. Among the Canidae there are many different ways of life and all members, while preferring to adopt that to which they are most suited, can in extremity find alternative ways of obtaining their food, security and other needs. Some examples are the following: hunting in packs; solitary hunting; carrion eating; scavenging; subsisting on lower forms of life, such as crabs or molluscs, insects, slugs, and small vermin such as rats and mice; eating vegetable foods, including leaves and fruits, and crops such as maize and roots. With the exception of pack hunting (which is practised only by some), all Canidae can and will at times do all these things. When rabbits became scarce as a result of myxomatosis, foxes lost a succulent dish of which they were very fond, but not their source of livelihood. Jackals too, although depending chiefly on scavenging and eating of carrion, also capture small animals and birds and, when pressed, will revert, in northern India at any rate, to pack hunting.

In their breeding habits also, virtually all Canidae share common characteristics in that they either dig burrows in which to bear their young and find refuge, or they find holes or retreats in tree trunks or among rocks. To a greater or lesser degree, all Canidae also have the instinct to store away food. This is developed especially among the arctic foxes, but is also seen in the habits of domestic dogs when they bury bones. All Canidae bear multiple young. This is a characteristic important to survival in a short-lived type of animal and it is inherent in the polymorphic genetical system.

This system can only provide change in relation to altered circumstances if, under the harsh laws of nature, excess young are produced and those born with less favourable characteristics are eliminated by processes of natural selection. We may suppose also that the production of excess young was important in the hierarchical system; social outcasts are driven to the borders of the pack, where they will fall victims first if the pack is attacked by predators. The main breeding core of the most suitable individuals will therefore be preserved and the least suitable will be sacrificed. That this happens also in other groups is well known, for example from the eminent ecologist Paul Errington's researches on musk-rats.

A pack of wild Canidae thus contains individuals of different conformation and temperament. The effect of selective breeding by man is to build new breeds in which certain inherent properties are intensified; but these new breeds no longer possess the great diversity of talent present in a wild pack. When dogs become feral the opposite process takes place: inbred characters are discarded and the members of the community revert to ancestral type both in relation to conformation and temperament.

The division of labour which has given rise to a social diversity in groups of Canidae is perhaps more intensified among these animals than in any others. This has evidently come into being in response to the necessity for a small predator to attack very large prey which as an individual it would not be able to kill on its own. The social organization of Canidae in respect

1. Caucasus dog. The picture is of a small pointer-type dog, photographed
by the author in the foothills of the Caucasus Mountains, near Sukhumi
in Georgia, U.S.S.R. The dog, which was common in this region,
bears obvious resemblances to Pointers and to Dalmatians. The Windmill Hill
dog skeleton also reveals a dog of this type. We regard this as representing
the prototype of the retriever/spaniel group of dogs, which today extends through
the mountain ranges from India to western Europe and has existed in these
areas since Neolithic times.

2. Assyrian mastiffs. During the era of Assyrian dominance in the Middle East – c. 600 BC – enormous dogs of mastiff type were bred for purposes of war and for hunting large animals such as wild horses and lions. These dogs were depicted in action engraved on stone friezes, from which the breed characteristics are clearly discerned.

3. Egyptian greyhound,
1250 BC. Dogs of greyhound
type were bred in Ancient
Egypt from very early times, as
is shown by this plate. They are
usually depicted with the
uncharacteristic coiled tails
shown here.

4. Actaeon devoured by his
hounds. Greek, second century
AD. Greyhounds became
favourite hunting dogs with the
Greeks – and indeed the name
is sometimes supposed to be a
derivative of 'Greek hound'.

5. Han Dynasty dog, 212 BC. Chinese dogs, from earliest times, appear to have been derived from multiple sources including Maltese, Mastiffs, Greyhounds and Dingo or Pariah type. This dog is plainly of mixed breed with Chow characteristics, and shows the pariah influence in the curled tail, which resembles that of the Samoyeds and Elkhounds.

6. The Fo dogs of China, symbolic of the lion of Buddha, were kept in many Chinese households portrayed in pottery, ivory, jade, and other materials. The pair depicted here, from originals in the possession of the authors, shows the traditional appearance of these dogs. The male has his right foot on a ball of bamboo, the female her left on her puppy. Sometimes the puppy is shown attempting to climb up her left leg. The fierce appearance is to frighten away evil spirits.

7. Chinese silk picture showing dogs of Pekingese type, eighteenth century. The Pekingese was bred in the Imperial Palaces of China for beauty of form and temperament over a thousand years or more. The result was a dog of bizarre form, but of great beauty and character, embellishing many notable works of art, such as that depicted here.

8. 'Lootie' was one of five Pekingese from the Imperial Palace in Peking brought back to England in 1861 by General, then Captain, Dunne, and presented to Queen Victoria. The dog was so small that it would curl up and go to sleep in the General's military cap.

9. Peruvian dog. Pottery vase of Nazca Culture, *c.* AD 100–900. Some early dogs of Peru and Mexico bear a close resemblance to the Ha Pa dogs of China, from which the short-nosed dogs of Pug type are derived. It is interesting to speculate whether the Mongoloid peoples of the American continent brought with them dogs similar to those taken to China by people of similar race, which were ancestral to the Ha Pa. The Mexican dogs of this type have contributed to the ancestry of the chihuahua.

10. Mexican hairless dog. Among other kinds of dog, a hairless breed was developed in Mexico, as with the Indian Polygar. A feature of this breed is the body temperature, which is chronically elevated above the normal level for other breeds of dog.

11. Basenji dog. The bitch 'Sula of the Congo' was born in the southern Sudan in 1959. She was brought to England when 3½ months old by her owner, Miss V. Tudor Williams. Many English and American champions are descended from her.

12. Dingo, the so-called Australian wild dog, which is both kept in captivity by the aborigines and lives and hunts as a wild animal. These dogs are believed to have been taken to Australia in late Palaeolithic times.

13. Mrs Connie Higgins' bitch 'Shebaba'. A feral breed of dingo type found in the Middle East. These dogs live in desert regions and hunt in packs. They have recently been brought to this country, where they are being bred by Mr A.S.Mold. They take readily to life in domestication.

14. New Guinea singing dog. These are small dogs of typically dingo type, domesticated by the natives of New Guinea. They are so-called because of their melodious howling usually to be heard at sunset.

15. Husky dog, the most primitive of the Northern Group of Dogs. These dogs are very similar to the Arctic wolves and are bred back to them at intervals by the Eskimos.

16. A wolf (arrowed) rounding up caribou. It is running into a herd of barren-ground caribou. Those in front of the wolf are running, those a little further away are walking, and those on the edge of the herd are lying down.

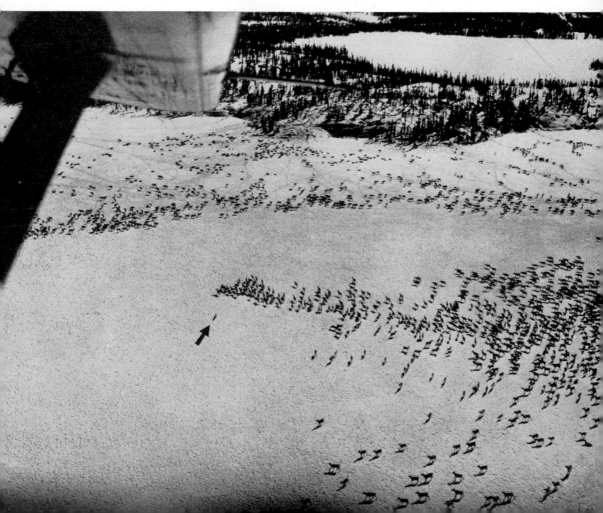

17. (*opposite*) A team of Huskies being worked by Eskimos in Greenland. European travellers would rarely use this formation and achieve better results with their dogs.

18. The odd dog out. Major Mike Banks, who supplied plates 17 and 18, informs us that there was one 'outcast' from the team depicted in the previous plate. This dog is shown here following the sledge team in a dejected manner. This illustrates how close these dogs are to the ancestral wolves, in whose social hierarchy there is always a bullied outcast.

19. The elkhounds show the typical characteristics of the Northern Group dogs. They have, however, curly tails and a shape and set of head suggesting admixture of pariah (dingo) blood, as seen also in the Samoyeds. This was introduced by the Finno-Ugrian peoples, who brought with them in their northern migrations dogs of pariah type, represented today by the Lapland Spitz, also now crossed with Northern blood.

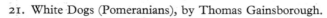

20. Pomeranian and Spaniel, by George Stubbs.

21. White Dogs (Pomeranians), by Thomas Gainsborough.

22. Greek greyhounds, male and
female, second century AD.
A mark round the neck of the
female hound indicated the
place of a metal collar.

23. A mongrel. The dog depicted here had a crossed boxer/labrador for mother
and a greyhound for father. His appearance reveals nothing of either
boxer or greyhound, and he is certainly not recognizable as a labrador,
though there are features suggestive of the 'curly-coated' retriever.

24. Northern wolf.

25. Maned wolf.

26. Coyote.

27. Side-striped jackal.

28. Black-backed jackal.

29. Common fox.

30. Raccoon dog.

31. Bat-eared fox.

32. Fennec.

33. Malayan wild dog.

34. Cape hunting dog.

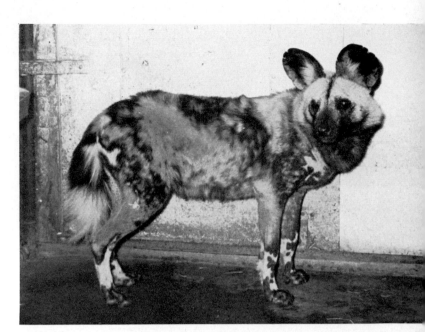

35. Bush dog.

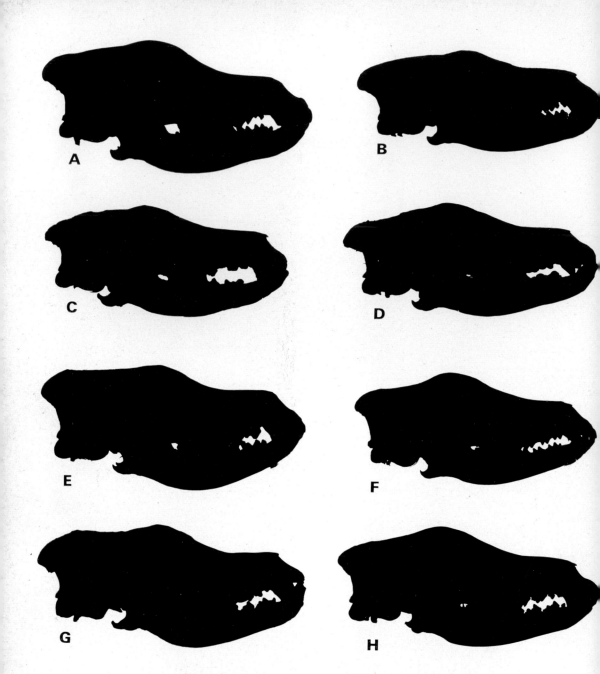

36. Skull silhouettes of Canidae. (A and B) Northern wolves. (A) is a Canadian timber wolf, and (B) is believed to be a Northern grey wolf crossed with a Timber wolf. Note extremes of variation in closely related wolf strains. (C and D) Skulls of (C) the Tibetan wolf (*Canis lupus chanco*) and (D) the Asian wolf (*Canis lupus pallipes*). (E and F) Skulls of (E) an Eskimo dog and (F) a dingo. (G and H) Skulls of (G) a Nepalese mastiff, for contrast with the Tibetan wolf, and (H) a collie representing the long-muzzled group of dogs; note the resemblance to the wolf in 36 (B), and to the Asian wolf (D) and dingo (F).

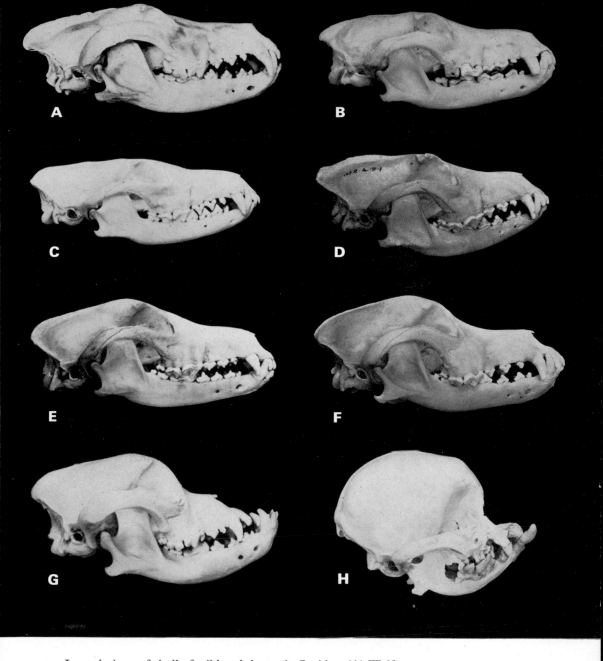

37. Lateral views of skull of wild and domestic Canidae. (A) Wolf.
(B) Coyote. (C) Fox. (D) Jackal. (E) Lurcher. (F) Golden Retriever.
(G) Boxer. (H) Pekingese. These figures illustrate how the differences between
domestic Canidae are greater than they are between wild Canidae, even those
so far removed from each other as wolves and foxes.

(a)

(b)

(c)

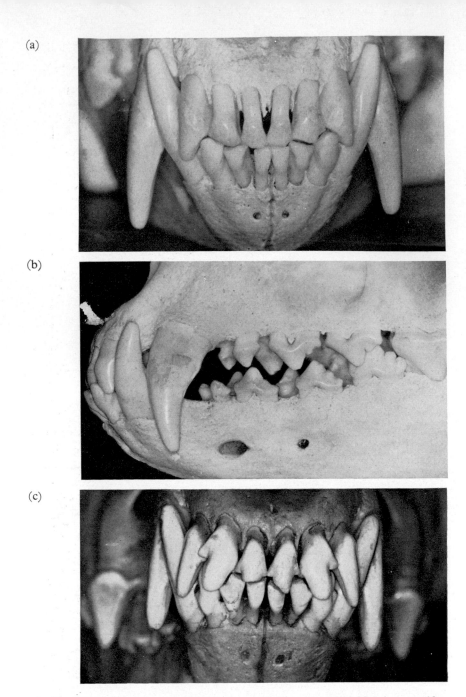

38. Teeth of wolves compared with domestic dogs. (a and b) Northern wolf, front and side views. The widely spaced powerful incisor and canine teeth are a feature of the wild wolf strain. The interlocking mechanism provided by the upper and lower canines and first upper incisor is a deadly weapon of offence, resembling a spring animal trap. This effect is enhanced by the reduction in size of the first premolars and the gap between them and the canines. (c and d) The front views of the tooth mechanism in (c) a pariah dog, and (d) an alsatian. The teeth of the pariah are well spaced and strong,

(d)

(e)

(f)

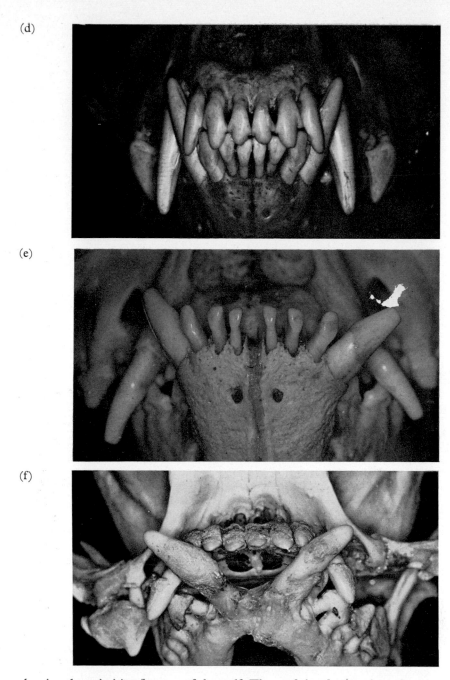

showing the primitive features of the wolf. Those of the alsatian show the
beginnings of narrowed jaws and crowded teeth, associated with domestication.
(e and f) The front views of the tooth mechanism in (e) a boxer, and (f) a
Pekingese, illustrating the extremes of jaw and tooth degeneration associated
with domestication and selective breeding for short muzzled characteristics.
In the lower jaw of the Pekingese, all the incisor teeth are missing
because of lack of development.

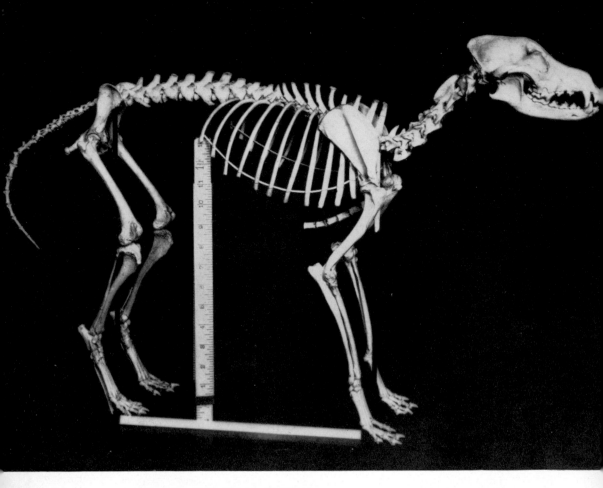

39. Skeleton of the Windmill Hill dog.

40. Pariah dog from the Temple of Denderah: (*left*) front view; (*right*) lateral view.

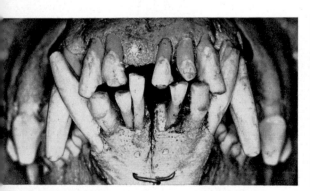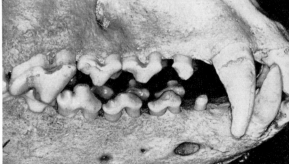

of contacts with each other is also particularly complex. In common with most other forms of wild life, territorial instincts are highly developed and adjoining families of wild Canidae scrupulously respect each other's territory, which is demarcated by scent signs introduced in the urine of the males. Recognition between individuals depends not on sight, but by distinguishing the characteristic odours present in the urine and in the secretions of the peri-anal glands, which are situated on either side of the anus and masked by the down-hanging tail. The highly developed territorial marking system and recognition displays are so obviously shown by all breeds of domestic dog that they do not need lengthy description. They have been studied and described by other authors such as Konrad Lorenz and R.H.Smythe.

It is difficult for members of the human race to appreciate that these scent signs are the dog's way of recognition, of demarcating home and property, and of finding a bitch ready to accept the male. We ourselves are so dependent on our sense of sight that we find it nearly incomprehensible that this sense plays little or no part in the social activities of dogs.

The psychological properties by which the Canidae support their way of life may be summarized as: intelligence; powers of sight; hearing; scent; sound and communication. No doubt it will be argued on theoretical grounds as to whether they do have intelligence in the sense that it is possessed by human beings. Given certain facts, can dogs assemble them, draw deductions, and base action on reason rather than instinct? Indeed can reason in any circumstances override instinct? It is evident to us that the more intelligent dogs at any rate do have such powers; we do not believe that this will be questioned by anyone who has intimate practical knowledge of dogs, or by those who have experience of working sheepdogs. We consider that in the course of pack hunting wild Canidae also exhibit powers of rational thought and planning and that this is an essential element in this type of hunting.

The dog's powers of sight, although unimportant in many spheres where they are used by man, are keen and powerful. It seems that scent-hunting hounds have well developed powers of sight, but they do not use them extensively; in the sight-hunting hounds, on the other hand, such powers must in the ancestral forms have been important in obtaining food and to survival generally. Powers of hearing are good and Canidae can hear sounds at ultrasonic levels which are inaudible to human ears; these powers are used for controlling dogs by means of ultrasonic whistles.

Primitive Canidae possess either prick ears, or ears which can be laid flat and raised at will. The external ears can be orientated in different directions, enabling the dog to determine the direction from which a sound comes; such a property is lacking in man, who finds it difficult to discern the direction of a sound without moving the whole head. It would appear that in primitive Canidae it may be important to orientate sound both on account of security and for detection of the direction in which prey lies.

Probably good hearing is also valuable to dogs when hunting, particularly in thick country, in order to locate and keep in communication with each other. The baying of scent-hunting hounds has a differential character with different meanings, easily discernible to the huntsman and presumably to other hounds in the pack. In addition, each hound has a characteristic bay, so that huntsmen can locate each individual, and no doubt the dogs themselves are also able to tell whether the sounds come from one of their number or from a stranger. As with many other aspects of hunting behaviour of wild Canidae, little is known of the extent to which the different groups use sound and hearing in communication, either for recognition of their own pack, or for combination in hunting, although it is accepted hat among the northern wolves the characteristic howl is used as a recognition signal.

The power of scent is all-important in canine activities in relation to territory and mutual recognition and in all groups it is also important in hunting. While we speak of the greyhound group as sight-hunting hounds, it must not be overlooked that scent is equally important to them. Once located, the prey is followed by sight with great speed; but scent is used to find the prey initially. Among those Canidae which find a living in close country in mountain or forest, as would be expected scent is all important both in locating and following prey.

Psychological patterns in wild Canidae conform closely to ways of life. The fox derives his living from finding small creatures by stealth and cunning, qualities which have been legendary from time immemorial. Jackals are great scavengers, often obtaining their food by slinking into human habitations and stealing what they can find. In temperament they are timid and alert, shrinking and cringing, and lack the boldness and ferocity of the true wolves. Their ears are large and their hearing acute, so that they are readily alerted if disturbed at their nocturnal activities. The northern wolves from which our spitz and sheepdogs are descended are well organized for pack hunting and seek their prey in either forest or tundra. In them is developed the herding instinct, by means of which herds of prey animals are cornered or driven to convenient locations and single animals are separated from the herd. They are intelligent, powerful and fierce. Both scent and sight are well developed and, it may be suggested, also a high degree of intelligence.

The mountain wolves from which the scent-hunting hounds are derived tend to be strong and massive and are able to tackle large animals. They have especially well developed powers of scent, by which they hunt animals in thick country where sight is of little use; it appears also that they have means of keeping in touch with each other by baying and other sounds. Prairie and desert wolves, such as coyotes and that form of wolf which gave rise to greyhounds, scent on the wind and hunt silently by sight. They are endowed with the stamina and speed of a cursorial animal and have the psychological and temperamental attributes which suit them to their particular habitat.

When domestic dogs become feral, as has happened with those of dingo type, they too revert to an ancestral form suited to the habitat. Pariahs, like those described earlier belonging to Mrs Connie Higgins and Mr Mold, have taken to hunting wild animals and have reverted to something resembling the ancestral dingo; those which scavenge around cities, however, become cur-like.

We have already discussed the special significance of tail formation, which remains curly in the scavenging pariah but which in the hunting pariah again becomes drooping. Masking of scent by the tail, which is plainly of great importance to wild hunting Canidae, is also seen in the domestic dog. In the happy, exhilarated dog the tail is held high; in the dog that is scolded it is lowered between the legs, thus masking his scent and promoting concealment.

Thus, the social attributes of wild Canidae, closely linked to their way of life, can readily be traced in the main groups of domestic dogs derived from each form of wild wolf. This lends strong support to our contention that the domestic dog has been derived from at least four different ancestral wolf stocks, which have contributed to their descendants the special characteristics they developed in response to the dictates of their habitat. The gregarious and social instincts of dogs which have rendered them acceptable companions to man have arisen from their adaptation to life as social predators. This has been possible only because of social behaviour patterns, and these patterns in their turn cannot exist in the absence of a wide polymorphic genetical system. This system has made possible a wide diversity of breeding potentialities and the great range of temperament possessed by domestic dogs. It has also made dogs economically desirable to man, because he was able to breed them for the multiple uses which were important to him in the past.

Among mammals, probably only dogs and man are truly social animals in the sense that they can combine and communicate to get their living. Wild dogs have come into competition with man: wolves and dingos are disappearing and the numbers of foxes are controlled. Yet as an ally the dog – selectively bred – maintains his position and anomalously is used to hunt and kill wild Canidae.

Unfortunately, unless specially trained, dogs are not naturally clean animals. Puppies must be house-trained and even adult dogs are a nuisance in towns because of the way in which they foul the pavements with their excrement. The territory-marking habits of the males in urinating against lamp- and door-posts are among their less endearing characteristics. Also, their everlasting use of the nose to investigate anything which stimulates their overflowing inquisitiveness leads them to contaminate this organ with all kinds of filth. Some dogs take great delight in rolling in all the muck they can find, acquiring odours offensive even to the most insensitive human nostrils. They are not adept at keeping their skins and coats clean and require frequent brushing, combing and bathing if they are to be fit companions.

This lack of hygienic instincts undoubtedly arises from their ancestry as roving animals which do not acquire a permanent abode. In these respects they provide a marked contrast with cats, whose devotion to a place rather than to a person is so characteristic. Cats keep their coats in immaculate condition and conceal their excreta; this inborn cleanliness even in very young kittens is supposed to be due to the habits of their ancestors in living for long periods in the same lair. It is typical of animal species that those which are nomadic lack natural instincts of hygiene, whereas those which remain attached to one dwelling are instinctively clean. Taking an extreme case, birds and monkeys, which live in the air or in the treetops, have no hygienic instincts, dropping their excrement to the ground below where it cannot serve to lay a trail for a predator. In general, it is impossible to train birds and monkeys to be clean.

Much of the seemingly unreasonable dislike of dogs among oriental people arises from an abhorrence of these dirty habits. One of us (R.N.F.) recalls a conversation he had in Kenya with a Somali friend of Mohammedan faith. It was explained that it was not the dog he must not touch, but its nose. Virtually this meant the dog, because you could not touch a dog without his investigating you with his nose (and indeed unless you let him sniff first the back of your hand, you are in danger of being bitten). The nose, he explained, was sure to have been inserted previously into all kinds of muck and was undoubtedly physically as well as religiously unclean. This was highly likely to have been the case in Middle Eastern cities, where in addition to all manner of refuse, including human excreta, even human corpses were left lying about or thrown on to the dung heap. This abhorrence of dogs, which was prevalent in Biblical times, was soundly based on hygienic concepts and arose from the unclean nature of the dogs themselves. A study of Leviticus, and Mosaic law generally, shows to what extent the control of disease by hygienic laws had become important to formerly nomadic peoples once they settled in cities.

Apart from the physical transfer of noxious organisms on portions of their anatomy, resulting from their unclean habits, dogs can acquire diseases due to helminth parasites which they can pass to human beings. One that causes concern is the round-worm larvae known as *Larva migrans*; these infect children and migrate through the body tissues, sometimes passing to vital organs and causing serious illness. If these larvae pass into the eyes they can cause blindness. Normally the effects are insignificant, but infestation of dogs with the parent adult worms may arise from their habits, the things they eat becoming contaminated with another dog's excreta. Another very serious disease, we have already mentioned, contracted by man from dogs is hydatid cyst, which comes from acquiring the larval form of the small dog tapeworm *Echinococcus granulosus*. This again results from the habits of dogs in passing their excreta in places or under conditions in which it contaminates the hands or food of human beings.

Such troubles become of less importance when dogs are kept under proper hygienic conditions, and in advanced countries they are not of sufficiently frequent occurrence to diminish the dog's popularity. Nevertheless, it is due to the defects of the dog's character that in some countries and under certain conditions he is regarded as an abomination instead of the ally which he was initially.

With regard to behaviour patterns generally, the dog is what he was when man first brought him into domestication. He had his qualities good and bad, which all played some part in adapting him to his environment and which endowed him with the capacity to survive in a competitive world. By selective breeding, man has developed some of them and suppressed others. Man cannot, however, create in him what did not exist originally. The outstanding feature in dogs is the diversity of those qualities; this has enabled man to produce breeds of dogs with such a variety not only of conformation, but also of temperament.

Many works deal with the significance of the various social behaviour characteristics of domestic dogs and a detailed survey of them is irrelevant here. In temperament, dogs are as variable as they are in conformation. They can be bred or trained to be fierce and unmanageable, or to be lovable, loyal, and faithful. Dogs possess inborn temperaments as diverse as those of man himself; they may be bounding, joyous creatures, they may be dour and phlegmatic, or they may be manifestly disagreeable and unpleasant. Thus they all have a basic pattern of behaviour derived from their ancestral ways of life; within this norm, they show extremes of ability and temperament.

Chapter 14

Genetics and Inheritance

WHAT is a dog? Obvious though this question may sound, the time has come when we must attempt to answer it scientifically. Even a child seeing a dog, whether a great dane or a chihuahua, instinctively recognizes the animal as a dog and can put a name to it in spite of the great disparity of size, colour, and conformation. Yet the question 'what is a dog?' is not altogether an easy one to answer. We can attempt to answer it scientifically by descriptions of features of the skeleton, the shape and number of the various bones. Yet unless we take account of minor features of tooth number, we shall include among the dogs genera such as the dholes, which are not generally regarded as true dogs. We shall certainly include jackal, wolf, and fox, although most people would deny that they are in fact dogs.

Alternatively we can say that we regard these various animals as dogs provided that they can interbreed and provided that the progeny are fertile when mated with each other. Dogs can breed with wolves and jackals and foxes and the progeny are indeed fertile; but many breeds of domestic dogs cannot interbreed because of their disparate size, making it physically impossible to mate. Thus this definition becomes invalid for the subject of our study, the domestic dog itself. We must then qualify our definition to mean animals that could interbreed if size and other differences did not prevent it, and we could probably prove that a great dane and a chihuahua could produce fertile offspring by means of artificial insemination.

The methods by which breeders produce new breeds are simple enough. Whether related or not, dogs having the desired characteristics arc mated; the pups from the litter are again selected and mated, and in a reasonably short time the characteristics sought will have been intensified, though possibly at the expense of other desirable features. It is here, however, that the difficulties arise. Such animals, if produced by outbreeding, will not necessarily breed true and only a proportion of the pups will have the desired features. A new breed has not, therefore, been established; to do this the progeny must be inbred so as to 'fix' the breed. This is normally done in one of two ways: either by mating selected brothers and sisters from the same litter; or by breeding back suitable progeny to dam or sire. By this means, the selected characters are reduplicated in the progeny and the breed is 'fixed', but so are any undesirable characteristics that go with them. The breeder gains something he desires, but at the same time he loses other properties which may be of importance, such as resistance to disease, stamina or intelligence. He may also be duplicating inherently bad features such as a tendency to congenital deformities or inherited diseases, as for instance cleft palate, haemophilia, or deafness.

The mating of brother with sister to fix the breed, while the most effective method, is also the most drastic and in general breeding back to sire or dam is preferable. Sometimes two desired characters prove to be incompatible except in cross-bred species and a breed cannot be fixed. In this case the breeder must maintain two breeds, crossing them with each other and using the crossbred progeny. This usually occurs where certain colour phases of show dogs are desired. Even with inbred canine stocks, there is a tendency for puppies to be born which are not true to type, so wide is the gene pool of dogs. The more developed breeds are maintained purely artificially, because under wild conditions they would not be able to acquire their food and survive. Much has been bred into them, but much has also been bred out of them.

Recognition of the extreme variation among breeds of domestic dogs should make it easy to understand how it is that within a single polymorphic genetical system such diverse animals as wolves, jackals and foxes still belong to the same genus and are still variations on the same genetical theme.[1] Those animals, such as the koala bear and the giant panda, which become greatly dependent on a single source of food or a single type of habitat, become progressively inbred. Their genetic pool becomes more confined; they breed more and more true to type; and more and more of them are homozygous rather than heterozygous. Animals which have a wide geographical distribution and can tolerate wide variations of climate, can subsist on a great range of foodstuffs, and become progressively more independent of the habitat. At the same time they widen their genetic pool and develop heterozygosity at the expense of homozygosity. They also have a greater power of variability within the stock without a change of the fundamental genetical constitution.

This concept can be reduced to extreme simplicity in the following way.

[1] One of us (R.N.F.) has written about the importance of the polymorphic genetical system in relation to man in *Man, Nature and Disease* [23].

Let us suppose that the wolf-like characters of Canidae are represented by the letter 'A' and the fox-like characters by the letter 'B'. Then a canid population could have the genetical composition AA, 2AB and BB, AA and BB being homozygous and AB being heterozygous.

Let us now consider that a habitat is suitable for wolf-like canids, but that fox-like canids cannot survive. Then all BB animals will fail to survive to adult life or will not reproduce, that is they are genetically unsound. Our population then consists of AA and AB, with a preponderance of dominant A genes in the ABs so that the population becomes wolf-like. Where the habitat can only support fox-like animals, the converse will hold true, so that the population then consists of BB and AB.

Let us now suppose that the climate changes so that the first habitat becomes favourable for fox-like animals. The AA types then disappear and BB types, derived from AB + AB matings, take their place. In a few generations the wolves change into foxes. Conversely in the second habitat the foxes change into wolves.

This is, of course, a gross over-simplification, since both A and B represent tens of thousands of interacting genetical factors. It does, however, demonstrate the importance of the heterozygotes in a polymorphic genetical system, and it illustrates how animals so diverse as the Canidae could all be variations of a single gene pool.

We have already commented on the benefits of outbreeding in dogs, comparing the popularity of those breeds derived from mastiff stock (and evidently containing a multiple ancestry) with those of dingo stock which do not have multiple capabilities. Breeders, of course, are well aware of the advantages of producing new breeds by crossing dissimilar canine stocks, and this explains the difficulties experienced over attempting to classify canine breeds. It is, however, noteworthy that even when canine stocks are crossbred, one or other of the parent strains appears to leave the greatest mark on the progeny. This is well shown in the picture of the crossbred dog in plate 23; it would be difficult to guess that this dog had a crossed boxer/labrador for its mother and a pure greyhound for its father. Its appearance indicates nothing of boxer or greyhound; nevertheless its progeny would probably segregate into boxer and labrador and greyhound types if he was mated with his full sister.

Apart from our comments on the polymorphic system, we have also suggested that the wild northern wolves have a particular dimorphic system. This suggestion has been advanced on the evidence of only a limited number of skulls; yet the two different conformations clearly resemble the two main groups into which domestic dogs, derived from northern and mountain wolves, have segregated. If these hypotheses are correct, much in the descent of different domestic species from wolves is easily explained. Thus we could consider that among the mountain wolves of Anatolia and Tibet the short-muzzled, massive-skulled types would predominate, and from them would come dogs of mastiff type. Among the Asian or African wolves, the lean-skulled, fine-drawn type would predominate, and from them could come the ancestral form which gave rise to the greyhound types.

The main features which are variable in dogs have been assessed and enumerated by some students of the subject. The ways in which these characters are transmitted by different types of crossing have been evaluated and much information is available. This has recently been summarized with great ability by S.A.Asdell [1] of Cornell University in his excellent little book entitled *Dog Breeding: Reproduction and Genetics*. It is not our purpose to delve deeply into this problem and that which follows is largely a summary of Asdell's review.

Work on the chromosomes of domestic and wild dogs presents a number of difficulties because of the very large number they possess and the complexities of counting and typing them. It is generally accepted, however, that *Canis familiaris* has 38 pairs of chromosomes, together with the X or female chromosome and the Y or male chromosome, making 78 in all. Professor Herre, of the Institute for Domestication at Kiel, has informed us that there is still some doubt as to the precise number of chromosomes in domestic dogs; it appears that it is the same number as in wolves but possibly jackals have one additional pair. Nevertheless, jackals can mate with dogs and produce fertile offspring in spite of the dissimilar number of chromosomes (if it exists).

Asdell [1] enumerates the main variable characters of dogs as follows:

1. those appertaining to the axial skeleton (the skeleton of the body);
2. those appertaining to the appendicular skeleton (the limbs);
3. the shape and set of the head;
4. the type, size and carriage of the tail;
5. the size and set of the ears;
6. depth or intensity of pigmentation;
7. the basic colour series;
8. colour patterns, i.e. white pattern, ticking and roan, colour intensity factors;
9. behaviour traits, fearfulness, excitability, hearing and touch, hunting traits;

10. hair types;

11. eye and skin colour (i.e. as opposed to coat colour);

12. blood types.

The combinations and permutations between these various basic characteristics can account for all the differences seen in domestic dogs; and the factors controlling them are carried on the genes in all of them. They are studied by crossbreeding dogs which carry the genes for the various factors and seeing how each characteristic is inherited. The genes for some factors are dominant, for others recessive, for others they determine intermediate characteristics. In some genes dominance is incomplete, and there are also modifying genes which may modify the effect of a gene in character determination. Some characters are determined by what are known as multiple alleles, that is to say more genes than one acting in community.

With artificial breeding of any animal, including man, it is inevitable that some abnormalities are produced and some of these, though lethal under wild conditions, will be preserved artificially. Some are the result of undesirable mutations and some are sex-linked. Thus haemophilia in dogs, as in man, produces symptoms of bleeding in the males, but the trait can be inherited through the females.

These matters are so well described by Asdell that we shall not pursue them further. Indeed, to do so would not be appropriate in a work on the natural history of the domestic dog. We note merely that these factors can, in various combinations, produce the great range of domestic dogs; that they are present in the ancestral species which man has used to select his breeds; that they are inherited according to Mendelian laws and well understood principles of genetics; and that breeding these animals by artificial methods in order to intensify the desired characteristics may lead to greater or lesser weaknesses, some of which may produce undesirable characters.

The termination of this chapter brings us to the point where we must summarize what has been learned from our survey. We have attempted to lead the reader by logical stages to an understanding of the animal under study, on the face of it an animal of bewildering complexity. Surely no other single species can show such diversity. Undoubtedly any wild animal having this extreme range of conformation of skull formation, tooth numbers and structure would be classified as falling into several species, if not genera. Yet not only are all our domestic dogs variations on a single theme and closely related to each other, but also the various other Canidae from which they are derived offer very little to the taxonomist by which he can separate them into species; and even when separate genera are recognized, they rest on rather flimsy bases.

The answer undoubtedly lies – let us repeat it – in the unusually wide polymorphism of the genetical system, which gives rise to a great range of heterozygous states (that is, combinations of diverse genes which on outbreeding easily segregate to produce apparently new conformations). Those two ancient allies, dog and man, share this same tendency, and to it must be attributed their ability to adapt themselves to a greater range of climate, food and other conditions than most other animals.

We regard the flexibility of the genetical system in both man and dog as inseparable from the achievement of social forms of living. Social life means the development of individuals with varying capabilities. With man, some are soldiers and sailors, some financiers, some shopkeepers, others electricians or labourers; each has his place in the community and without these varied skills, there could be no community as we know it. In a lesser way, the wild canine species have developed a great power to segregate, so that individuals even from the same litter may be leaders or subordinates, while others may even be outcasts, bullied and driven to live on the out-skirts of the pack. These outcast wolves are the ones most exposed to predation; so in early human communities, individuals were sacrificed to the gods that their death might appease the predator and save the com-munity. Only in man and dog among Mammalia has a social organization of such complexity been developed, though it exists to a lesser extent among other primates, particularly among those which live on the ground, like baboons. Social organizations comparable to these, however, are well known among some groups of insects, notably bees, ants, and termites. In them too social organization has led to the production of different forms of these creatures, though the way they are produced is quite dissimilar.

Quite the most intriguing suggestion which has emerged from our studies is that the wild northern wolves, apart from their polymorphic traits connected with abilities, show a special dimorphism associated with conformation and build. This concept shows that the traits necessary to produce the fine, long-nosed, highly intelligent breeds of dog were already present in the wild wolf stock, together with those which tend to produce the short-nosed, scent-hunting dogs with great abilities as warriors. The existence of a dimorphism of this type could well account for the ease by which Neolithic man had already developed the prototypes of modern canine breeds at so early a date.

Chapter 15

The Present and the Future

TODAY, man and wolf are ecologically incompatible. The depredations of wild Canidae on wildlife and domestic stocks are not tolerated; wolves are exterminated and the numbers of other Canidae controlled. The future of the canine race lies with the domestic dog, *Canis familiaris*.

The success of the dog as a partner of man in new environmental circumstances depends on two factors. The first of these is the hybrid vigour of some canine stocks, by which it has been possible to select them for varied and useful tasks. The second lies in the dog's possession of properties which are complementary to man's own.

By himself, man has limited powers of attack and defence. He has limited speed and endurance and his powers of sight, hearing, and particularly scent, are poorly developed in comparison with those of other predators, especially dogs. Man cannot detect the presence of game in thick country, nor can he readily follow and catch it. A man in unknown territory can easily become lost and be unable to retrace his steps. He must rely on organization, and the power of manufactured weapons and skill in their use. In the wolf he found to his hand a new 'tool' which he could adapt and which possessed those physical properties which he lacked.

Even today in many parts of the world and in connection with various activities the dog is of great importance to man, although in certain spheres of life this importance is diminishing. Yet never before were dogs so much in demand or so popular; the reasons are very different from those for which they were first domesticated, but they are such as to reinforce the belief that dogs still have a great part to play in human affairs.

Looking over his shoulder, man can still get a glimpse of the wide open snowscape of his ancestors. His path has led him through the forests of Mesolithic times, feared and propitiated by offerings to the spirits and gods of the woods, an age in which the dog was an indispensable ally. Today the path is narrow and misty, albeit with a distant glimpse of the stars to which man aspires. There still the dog star heralds the days of torrid heat and decrees the flooding of the Nile. Still Orion, accoutred with belt and sword, guides his dog across the heavens in hopeless pursuit of the rain stars, the Pleiades.

What do the stars foretell? Man is no longer guided by the simple beliefs of his ancestors. He attempts to master the cosmos, but does not submit to it. But cosmic forces may renew the glaciation or promote biological competition and bring to naught man's new-found science. In a materialistic age, attended by plenty and comfort unparalleled in history, and with knowledge and control of diseases as never before, there is greater perplexity, loneliness and frustration than has hitherto been known. Many people find the complexity of the more advanced ways of life beyond endurance. Simple pleasures no longer satisfy; competition between individuals and between peoples generates widespread tension and results in bizarre patterns of behaviour.

Some say: 'The more I see of my fellow men, the more I like my dog.' The dog still gives unquestioning love and devotion. He brings comfort to those who are distressed, companionship to the lonely. He retains within himself the old simple joys and love of life, the loss of which man himself perhaps subconsciously regrets. So many people have become dependent on the love and companionship of dogs that breeding them has become a major industry.

Inevitably, these newer needs dictate the breeding of dogs of different temperaments. The old forms persist, but the animal's character is changed. A show retriever may be of little use for retrieving, and a second breed of gundog retrievers must also be bred and maintained. Thus a new era is dawning in the breeding of dogs, whose main function now is to give companionship and pleasure. They have been bred for these purposes since before the dawn of history, but the need for them today is immeasurably increased.

The dog is still a contributor to man's newer way of life; and Cerberus still keeps many from the hell of loneliness.

Appendix I

The Genera and Species of Wild Canidae

TRUE wolves, classified as *Canis lupus*, include a large number of varieties indicated by subspecific names. In addition there are the coyotes, *Canis latrans*, and the red wolf of Texas, *C. niger*, the maned wolf of South America (*C. jubatus*), and the Abyssinian wolf (*C. simensis*).

Canis lupus (*Northern Wolf*)

The northern wolves frequent both forests and open country and may be seen either by night or day, singly, in pairs or in packs. They combine together in packs for hunting, especially in the winter when food is short. Singly, they will destroy sheep, goats or even children. In packs they will successfully attack animals as large as horses and cattle. If pressed, however, they eat birds, mice, frogs or any other small animals. At times they will feed also on carrion and even seek nourishment from buds and lichens. In their feeding habits, therefore, they grade into jackals, just as the jackals grade into the wolves.

They communicate with each other mainly with a loud howl. The males fight for the females in January and those that are successful remain with their mate until the young are well advanced in growth. The young are suckled for two months, following a gestation period of 63 days. The females make nests in burrows excavated in small caves or in dense thickets. Wolves come on heat only once a year, like dingos, but unlike domestic dogs (in their case the twice yearly heat period seems to be a response to domesticated conditions). Young wolves are full-grown at the age of three. They are easily tamed when young, but can never be tamed when captured as adults. When reared in captivity, they will interbreed with dogs and can learn to bark.

Wolves were exterminated in England during the reign of King Henry II, but survived in Ireland until around the year 1710, and in Scotland until 1743. The range of wolves today is more restricted than in the past, and they are being progressively eliminated from closely settled areas. This means that their numbers are much diminished, and they live largely in northern sub-arctic areas where conditions still exist suitable to their way of life.

Canis lupus pallipes (*Asian Wolf*)

C. l. pallipes is smaller and slighter than the northern wolf and has a shorter coat, with little or no under-fur. There are, however, no satisfactory distinctive characters by which the Asian wolf could be placed in a separate species. These wolves are confined to the plains south of the Himalayas, but in classical times were commonly found as far west as Syria and Palestine. It is said that they have reappeared in

Palestine since reafforestation and reclamation were undertaken by the government of Israel. It is supposed that they somehow survived in wild country during the period since Roman times and that their numbers are again on the increase. This statement is confirmed by Mrs Sonia Cole, who saw one in Syria in 1965.

Asian wolves combine in large packs less readily than the northern wolves, though six or eight will join together to seek their prey. In India they have developed undesirable habits and a large number of children used to be carried off every year, thus giving rise to stories of male infants being suckled and reared by wolves. Such stories are of doubtful authenticity. It is said that these wolves occasionally bark, but normally they howl, like the northern wolves. Breeding times are from October to December, mostly in the latter month. Cubs are born blind and with drooping ears.

This wolf is remarkable for its great speed and power of endurance. It is distinguished by its slenderness and in appearance it greatly resembles the dingo, though it is rather larger. Its slenderness, speed and endurance might possibly suggest that it has played a part in the ancestry of the greyhound group.

Canis lupus hodophylax (Japanese Wolf)

Japanese wolves are very like the common northern wolves, but they are smaller, the legs are shorter and the muzzles longer. They are the only wolves to be found in Japan and are clearly an offshoot of the northern wolves; they do not represent a separate species.

Canis lupus chanco (Tibetan Wolf)

The Tibetan wolves are a branch of the northern wolves, from which they are differentiated by slight differences in coat colour. It is said that in general the muzzles tend to be shorter than those of the northern wolves and this would make them suitable candidates for ancestry of the mastiff group. It is doubtful whether such differences can be established except as a trend, but nevertheless it seems probable that they have entered into the ancestry of the mastiffs, very likely as the main component.

Canis simensis (Abyssinian Wolf)

The Abyssinian wolf is little known, being confined to a very restricted habitat in the highlands of Ethiopia where few people have seen it. These wolves, often referred to as dogs, hunt in packs, preying on sheep and small wild animals, though they are not supposed to be dangerous to man. In size they resemble a large sheepdog, being definitely smaller than the northern wolves. Characteristically they have very long, slender muzzles and their colour is light yellowish to reddish-brown. The fourth upper premolar tooth (the carnassial) is much smaller in comparison with the upper molars, in contrast to the relative sizes of these teeth in other species of wolf.

The range of this interesting group of wolves or dogs is now very restricted, but this animal must undoubtedly have been known to the ancient Egyptians and in those days its range must have been much greater than now. It is not recorded whether these animals hunt largely by sight or by scent, though the latter would seem more probable from its forested habitat in the mountains.

Canis jubatus (Maned Wolf)

Maned wolves inhabit Paraguay and adjacent areas. They are the largest members of the canine family in South America, where they inhabit low, moist country.

They have exceptionally long limbs, large ears, and conspicuous colouration. Their muzzle is not much elongated and the fourth upper premolar (carnassial) is very short.

These wolves are not dangerous and they never attack man. They are of solitary habit and are said never to hunt in packs. However, the closely related long-legged wolf of Texas has very different habits. The maned wolves often pursue wild deer, but do not prey on domestic herds, although very occasionally they attack sheep. They eat pacas, agoutis, rats, birds, reptiles, and insects and they also feed on nuts, sugar cane and fruit. Their habits are largely nocturnal. They can be tamed and will mate with domestic dogs; the mongrels are said to be especially good animals for the chase.

Canis lupus antarcticus (Antarctic Wolf)

The Antarctic wolf, of which the numbers are rapidly diminishing, is a small species found only in the Falkland Islands. They do not hunt in packs, are not nocturnal, and are silent except during the breeding season. They feed largely on native geese. They burrow in the ground like foxes and pieces of seal and penguin have been seen at the mouth of these holes, suggesting that they either attack these creatures or consume them as carrion. It seems that they live mostly on rather soft foods, because the sagittal crests on the cranium are generally flattened, indicating a less powerful temporal muscle for mastication than is found in other groups of wolves.

These wolves greatly alarmed former visitors to the Falkland Islands, because they entered the sea and displayed great ferocity when the ships' boats put ashore. Probably they were quite harmless, but the sailors trying to land did not think so.

Canis latrans (Prairie Wolf or Coyote)

Coyotes range from latitude 55°N through Central America to Costa Rica and are especially abundant in northern Mexico, New Mexico and Texas. They hunt in packs and have a particularly bad name because of their howling at night, which gives a feeling of insecurity to travellers. Yet in spite of this reputation and the fear in which they are held, they do not normally attack human beings. These little wolves feed greedily on all kinds of animals, particularly rabbits, rats and young birds. They will also eat vegetable foods when necessary. In the autumn they are said to feed exclusively on the fruit of the prickly pear, and in winter they eat juniper berries. Coyotes are extremely cunning; traps are frequently set for them, but they quickly learn to avoid them.

Coyotes breed in retreats among rocks or in underground burrows. The young are born in May and June, in litters of five or six. They interbreed readily with domestic dogs and the crosses are fertile. There are no distinctive characters of skeleton or teeth which distinguish them from wolves or dogs and indeed, as we have seen, one breed of domestic dog owned by Amerindian tribes is derived from them.

Canis niger seu rufus (Red Wolf of Texas)

The red wolf of Texas is not described by Mivart and is also excluded from Pocock's *The Races of Canis lupus*, though there are descriptions in Young and Goldman's *The Wolves of North America*. These authors divide the species into three subspecies: *C. niger niger*, *C. niger gregoryi*, and *C. niger rufus*. The normal phase is 'cinnamon buff', 'cinnamon' or 'tawny' with grey and black. In the black phase, the upper parts of the limbs and tail are black or brownish-black. Wholly black specimens are often found in a litter with ordinary grey brothers and sisters, showing that the colour variation is merely a phase and it does not appear to be seasonal.

The subspecies *rufus* is found in central Texas and Oklahoma. It is very small and resembles the coyote, *C. latrans*, whose range it overlaps; indeed some specimens are difficult to distinguish. The distribution of the species as a whole was once around the Mississippi River valley and its affluents, south through southern Missouri, eastern Oklahoma, and doubtless western Kentucky and western Tennessee to the Gulf coast in Louisiana, also to the Atlantic in Georgia and Florida. Today it is restricted to the Ozark Mountain region in Missouri, Arkansas and south eastern Oklahoma, with a few outlying sections in Louisiana and Texas.

These wolves combine in packs, sometimes of very large numbers, to attack cattle or horses, but their chief depredations seem to be on sheep. In their general character and way of life, these wolves do not differ from the grey northern wolf.

The Jackals

Jackals have no distinctive anatomical features which separate them from wolves or domestic dogs. The sagittal crest, which is always present in wolves, may be present and well-marked, or it may be absent as in domestic dogs.

Jackals lack the savage defensive powers of the true wolves and are much more timid (they are preyed on particularly by leopards). Apart from this, the descriptions could apply equally to any of the small wolves such as the coyotes and it is quite evident that jackals are very little removed from wolves. Like foxes, they breed in burrows, which are usually more extensive than those of wolves, producing four pups in a litter. They will interbreed freely with domestic dogs and the offspring are fertile.

Canis aureus (Indian Jackal)

The Indian jackal has a much wider range than the Indian wolf, extending throughout India, Ceylon, Burma and Tegu. These jackals live in forests and open plains from sea level up to 3,000 or 4,000 feet. They make their appearance in populous cities and their omnivorous habits make them useful as scavengers. Occasionally they seize fowls or other small domestic animals. Outside towns, they will eat any animal they can subdue, usually hunting singly or in pairs. Sometimes, however, they hunt in packs, especially at night when they set up a great howling. Sickly sheep and goats fall a ready prey to these jackals, as well as antelopes which have been lamed or wounded. In default of meat, they will eat fruit or sugar cane, tree berries and ripe coffee berries.

They are often hunted in the same way as foxes in Britain. They are easily overtaken and pulled down by greyhounds, but have sufficient speed to give a good run to foxhounds. The European jackal is probably merely an offshoot of the Indian jackal; it is found in Greece and Turkey, as far west as Dalmatia, and also in the Caucasus and Asia Minor.

Canis anthus (North African Jackal)

The habits and mode of life of the North African jackals are similar to those of *Canis aureus*, though these animals are generally larger than those of India and Europe. There are slight differences in the skull measurements, which tend more towards the true wolf pattern. One skull from the British Museum (Natural History) examined by the authors and illustrated in plate 37 closely resembles the skull of a small wolf and could indeed be almost mistaken for that of a coyote.

Canis mesomelas (Black-backed Jackal)

The black-backed jackal is widely spread over southern Africa, including Ethiopia. In some respects it is intermediate between the Indian and North African jackal. The differences between jackal species are almost entirely those of colouration and it is hard to see why they should be assigned to separate species.

Canis adustus (Side-striped Jackal)

This jackal is found in the Kilimanjaro area. The snout is long and slender; the ears are not so long as in *C. mesomelas*, but longer than in the other jackals. The skull is remarkable for the length of the palate, which extends backwards beyond a line joining the posterior origin of the hinder true molars.

In the light of what was said at the beginning of chapter 14 about the variability of the genus *Canis* within a polymorphic genetical system, it is difficult to exclude the jackals from the common pool of Canidae from which the domestic dog might well have been derived. Our survey has revealed a number of qualities in jackals which at one extreme approximate them to wolves, just as *mutatis mutandis* at the other end of the scale the wolves tend towards the jackals. Thus jackals will combine in packs to hunt and will pull down quite large animals, giving tongue in wolf-like howls. At the other extreme, wolves – even the northern wolves – will scavenge as do jackals and will eat vegetable foods, insects or whatever they can find. There is therefore no essential difference between wolves and jackals; the one group has found itself in a type of habitat where its main livelihood can be secured by combining in packs to hunt, the other where its livelihood is more easily earned by scavenging and finding items of food that are less hard to come by. One wonders whether jackals are not descendants of those wolves which, perhaps a long time back in Palaeolithic times, followed nomadic human tribes, scavenging the offal left by man, and in later Mesolithic times playing a useful part in cleaning his settlements.

We have seen that for an animal of its size, the jackal has great powers of speed and endurance – not so great as those of greyhounds, but approximately equal to those of foxhounds. The conformation of the skull, the build, and the teeth make the jackal a promising candidate to have entered into the ancestry of some canine groups to which jackals could have contributed powers of sight and hearing. We have already commented on the probability that the ancient Egyptians, to whom jackals were sacred animals, crossed them with dogs. Those indefatigable breeders of animals would hardly have failed to experiment with such crosses.

Wild Dogs

A number of species of so-called wild dogs exist in different parts of the world. Some of these have been domesticated, but they are not important in relation to the ancestry of the domestic dogs as we know them in western countries. However, in order to complete the picture a rapid review of the various species will be given.

Canis magellanicus (*Magellanic Dog or Colpeo*)

This wild dog is similar to *Canis antarcticus* and by being classified as a wild dog rather than a wolf, illustrates the absurdity of these distinctions. They are found in a wide variety of terrain, from the humid forests of Tierra del Fuego to the almost desert country of northern Chile, extending along the western coast of South America for some 1,600 miles, but they live mainly in woods. They are chiefly nocturnal, but are sometimes seen by day. They have a rather weak voice, and are unlike other wild Canidae in that they bark like a dog. They are very strong and fleet but are not feared by humans. Their muzzle is much elongated and they have a distinct sagittal ridge.

Canis cancrivorus (*Carasissi or Crab-eating Dog*)

These dogs range through the forests and bushy plains of South America from the Orinoco to La Plata, but they do not extend into the pampas regions. They are commonly spoken of as foxes, to which they bear a resemblance; this again illustrates the absurdity of making distinctions between different groups of animals which merge imperceptibly into one another and which anatomically are variations of one and the same group. Sometimes these dogs attain a considerable size; they have an obtuse muzzle, a rather short tail and a powerful frame. They prey on small mammals like agoutis and pacas, and on birds, and sometimes they will combine in packs and run down deer. Their name, however, arises from the fact that they frequently catch and eat crayfish in the rivers.

Canis microtis (*Small-eared Dog*)

This small dog standing 14 inches high at the shoulder is found in Amazonia, frequenting the banks of rivers. The nose is elongated and pointed, the ears very short, and the fur also short and generally of a dark grey colour. Little seems to be known about its habits.

Canis azarae (*azara's Dog*)

This dog is found over the greater part of South America east of the Andes, but does occur on both sides of the mountain chain. It inhabits bushy regions, whence it makes excursions into the great forest on the one side and into the open country on the other. It seeks its prey at twilight and at night, feeding on small quadrupeds and birds. It will also eat frogs and lizards, and it bites through and sucks sugar cane in the plantations, being very wasteful and doing much damage.

The Azara's dogs hunt with nose to the ground like a hound, occasionally raising the head to the wind. In summer and autumn they go around in a solitary manner, but in winter the sexes associate and their loud cries are heard at night and in the evenings. Sometimes their cries are heard at other seasons, especially when there is a change of weather.

The males and females inhabit the same nest, usually in desert scrub, under the roots of trees, and sometimes in the abandoned burrows of armadillos. They do not make earths like other dogs and foxes. In the spring, the females bear three to five young; they rarely leave the nest during the first week, when the mother is fed by the male. When the young are weaned, both parents go hunting and bring back food for the pups. Towards the end of December the male leaves the family and the young follow the mother until she too leaves them.

When taken young, the puppies are very easily tamed; they know their master, come when called and even seek him themselves and lick his hand. They are not very obedient, however, unless a stick is used, and they have a very disagreeable odour. They are friendly and play with other dogs, but bark and growl at strange dogs. They sleep most of the day, waking at night to eat and play; often they go off at night, returning in the morning. They readily hunt with other dogs, but if the hunt lasts for several hours they get tired and go home.

Canis parvidens (Small-toothed Dog)

The small-toothed dog is an inhabitant of Brazil. It is remarkable for the very small size of the fourth upper premolar (carnassial).

Canis eurostictis (Striped-tail Dog)

The striped-tail dog is another Brazilian form much resembling *C. parvidens*. It has, however, an even more remarkable dentition, the first upper molar being exceedingly large.

Canis virginianus (Colishé)

This dog is also known as the 'grey fox' or 'Virginian fox', although it differs from the true fox and has affinities with South American Canidae. Its habitat extends from Virginia to Texas, Guatemala, Honduras, Costa Rica, and also Pennsylvania and Yucatan. It is much less enterprising and sagacious than the common fox and is often caught in steel traps. It is also less destructive to the farmer; it does not enter farmyards, although it may seek poultry which stray to forest verges. It feeds on any birds it can obtain, as well as their eggs; it will also catch small animals such as rabbits, the cotton rat, Florida rat, and voles. It also eats insects and vegetable food, especially ears of maize. Although mainly nocturnal, it is often to be seen in broad daylight; as a rule, however, it hides during the day in bushes or tall grass. Its howl is something like a coyote's but less abrupt, so that it cannot be described as a bark.

Sometimes these dogs will climb trees to avoid danger, especially when hunted with hounds. Their odour is less penetrating than that of the European fox, but they give good sport to huntsmen and a two hours' chase is usually necessary to capture them. Next to deer hunting, the chase of the colishé is the favourite sport of the southern states of the USA.

The females produce three to four young in a litter, between the middle of March and April in Carolina, but later further north. These animals make their homes in caves, fissures in rocks, or holes in the ground. They have certain unusual characters of skull and teeth which need not be discussed in detail.

The Foxes

Canis vulpes (Common Fox)

Foxes show as much variability as do the wolves. Even English varieties differ so much that they have been given different names, e.g. 'greyhound' (mountain fox), 'bush fox' or 'cur fox'. The differences lie in the pelage, absolute size, and relative proportions of various parts of the body. The total lengths of head and body of English foxes may differ to a very large extent, expressed by some observers as 100 to 170. Length of tail and ears is much less variable.

Foxes occasionally make use of burrows made by badgers or rabbits, though usually they excavate their own earths. They usually rest away from the earth,

APPENDIX I

either in woods or under the shelter of banks or hedges. The young are born in April, after a gestation period of 60 to 64 days, and there are usually four to six cubs in a litter.

Foxes prey on poultry, partridges, pheasants, hares, rabbits, rats, mice, moles, frogs, lizards, and eggs. They will eat all sorts of other things, such as cheese and butter when they can obtain it, worms, beetles, fish, molluscs, and other sea food such as crabs left on the beach by the tide. They will also consume carrion and vegetable foods, especially fruit.

As is well known, foxes are extremely wily, especially when hunted. They have a peculiar penetrating odour by which scent-hunting hounds can easily follow them. They become adult at 1½ years and live (if lucky) for 13 to 14 years. Dogs and foxes can mate and the offspring are fertile, though there is much more reluctance on the part of dogs than in the case of mating with other Canidae, possibly because of some repellent factor in the fox's odour. Foxes make a variety of sounds, ranging from yelps to barks and screams, and when at rest they emit a gentle murmur.

In general, as we have often repeated, there are no anatomical features which can distinguish foxes from dogs or wolves. Foxes do, however, have a character-istically long, sharp and very pointed muzzle and a very long bushy tail (brush), which is more or less cylindrical for most of its length. The eyes are oblique and the pupils nearly linear when exposed to strong light.

Fox species have the most extensive range of any Canidae. Unlike the wolves, they are found in Africa north of the Sahara, as well as in central and southern Africa. They range all over Europe and Asia to some distance south of the Himal-ayas and as far east as Japan. In America, they are distributed from the far north, on the shores of Hudson Bay and Labrador, and south to the latitude of north Mexico.

Canis velox (Kit Fox)

The kit fox is one of the most elegant and attractive members of the Canidae and the smallest found in North America. Formerly it lived on the open plains between Saskatchewan and Missouri and on the plains of Columbia. It burrows in the earth in country destitute of trees and bushes and moves very quickly; it is quite unknown in forest regions. In its tooth structure it resembles the red fox.

Canis lagopus (Arctic Fox)

The Arctic fox is definitely a distinct species, on account of external form, colouration and change of hue, peculiarities of cranial conformation, lack of odour, and habits. These foxes range all through the Arctic regions.

In summer the colour is a bluish or brownish-grey, but in winter some individuals become entirely white. These foxes are the only Canidae which change colour, in the same way as the ermine and variable hare. Sometimes both white and grey cubs are found in the same litter and it seems that there may be two dimorphic types – those which change colour and those which do not. It is said that there are some which are always white, some always grey, and some which change colour.

There are no characters of skull or teeth which differentiate the Arctic form as a separate genus from other foxes, jackals or wolves, though the head is less pointed than that of the common fox and the muzzle has a somewhat swollen appearance. Arctic foxes do not breed in solitary fashion like the red fox, but in little 'villages' of 20 to 30 burrows constructed next to each other. Towards the middle of winter they migrate southwards; and in some areas they are local migrants, unlike other Canidae.

166

Arctic foxes are easily tamed. They are very clean and do not foul places where they eat and sleep, and they have no unpleasant odour. They are less cunning than the red fox and they have a yapping bark These animals have the instinct of making hoards of food, consisting mainly of lemmings and hares. Living as they do in cold regions, the snow and ice give them a ready-made refrigerator and their hidden hoards keep in good condition. When given food, even when extremely hungry, their immediate instinct is to hide it rather than eat it. They will follow polar bears and feed on their leavings of seals and fish.

Canis corsac (Corsac Fox)

Corsac foxes are inhabitants of open country in central Asia; they avoid the forest-clad mountains of eastern Siberia, which are near their range. They are seldom seen by day, when they sleep in deserted marmot burrows. They often wander from one burrow to another, making no permanent home for themselves. This habit is a disadvantage to them, because their tracks can be seen across the snow and they can easily be snared. Old foxes are evidently aware of their danger and are very reluctant to move under such conditions; they have even been known to stay in burrows so long that they have starved to death rather than move out into danger. Their prey consists of small animals, like the Alpine hare and vole, and birds. They are very suspicious and savage and quite untameable, even when young. They are smaller than the common fox and the pupils of the eyes are round. They have a rank odour quite different to that of *C. lagopus*.

Canis ferrilatus (Tibetan Fox)

The Tibetan fox is probably a local variety of *C. corsac*, though the latter is long-eared and *C. ferrilatus* is short-eared.

Canis leucopus (Desert Fox)

The desert fox of India lives in open country, like *C. corsac* which it resembles. However, *C. leucopus* is a rather handsome animal, with distinct colouration. Its range extends from the Punjab and Rajputana through Afghanistan and Iran to Arabia and Egypt. It lives chiefly on gerbils, which are very abundant in the sandy regions which it inhabits. This fox is a rapid runner and gives huntsmen good sport even with English dogs.

Canis bengalensis (Bengal Fox)

The Bengal fox is common throughout India except in forested regions. It is a small, very attractive and elegant animal, with slender limbs, short muzzle and a bushy tail. It is readily tamed and is a playful, frolicsome and agreeable pet, being very clean and having no unpleasant smell. It is often seen, being not at all shy, and frequently enters gardens and other enclosed spaces. However, it rarely takes poultry or indulges in other undesirable activities. It utters a short yelp, quickly repeated three or four times – a sort of little chattering bark.

These foxes pair from November to January and breed in burrows, usually producing four young in the litter, born in March or April. The burrows always have several branches radiating from the centre and opening separately; they are made in open plains, sometimes in thorny scrub, and these foxes also live sometimes in cavities of old trees. They feed on lizards, rats, crabs, white ants and other insects and are very adept at catching moths. They hunt quail and eat young birds and eggs, and also habitually take vegetable food, such as melon and other fruits, pods, and shoots of growing plants.

They are much coursed in India and give excellent sport with locally bred dogs, although they are too slow to escape from English dogs. They have very poor scent and cannot be hunted with foxhounds.

Canis canus (Hoary Fox)

These very small foxes of south-west Asia are delicate little animals and remarkable only because they have a peculiar feature of the skull in the very prominent auditory bullae.

Canis procyonoides (Raccoon-like Dog)

This strange animal, which is usually classed with the foxes, is an inhabitant of Japan and parts of China. Its flesh is much relished in Japan and its calcined bones are valued as medicine. The raccoon-like dogs live in woods on mountain slopes and are said to climb trees to get at fruit; they are not destructive to poultry. They rest in hollow tree trunks and burrows and are of exceptional interest because, unlike all other Canidae, under certain conditions they hibernate like badgers. They only do this if they are fat and in very good condition, in which case they will go to sleep during the winter in a deep burrow; sometimes the burrow of a fox is chosen, but it must be deep so as to go below frost level. If not hibernating, in winter they frequent running steams to feed on fish. These are the most omnivorous of all foxes, feeding habitually on vegetable substances, largely acorns. They are especially fond of fish and if given the choice of fish or meat, will eat the fish first.

In captivity they become accustomed to human masters, but they are very timid and not at all savage; they are also very clean. Their movements are somewhat civet-like and they sleep in a peculiar manner, rolled up in a ball. Unlike other dogs, the dorsum of the skull is strongly concave; the muzzle is very pointed.

Canis chania (Asse Fox)

This long-eared fox of South Africa tends towards the type of the true fennec, but is much larger. It inhabits Namaqualand, on both sides of the Orange River.

Canis pallidus (Pale Fox)

The pale fox is a long-eared animal, like the Asse fox but smaller. Its habitat is East and West Africa. Unlike most foxes, it has a characteristically short muzzle.

Canis famelicus (Ruppel's Fennec)

This little fox from the Nubian desert also has very long ears like the true fennec.

Canis zerda (True Fennec)

This beautiful little animal is recognizable by its extremely long ears. It is a desert fox, inhabiting North Africa from Nubia to Algiers and is found throughout the Sahara. In captivity it is very tame and gentle. It eats dates and any sweet fruit, is very fond of eggs, and it probably catches birds and small animals. It has very large auditory bullae.

The Dholes

The genus *Cyon* comprises the dholes or Indian 'wild dogs', but these are never-theless generically distinct from dogs. They are to be found from Siberia to Java.

Cyon javanicus (*Southern Dhole*)

The southern dholes are larger than jackals, but vary in size as well as colour. They have moderately long tails, which may or may not be bushy. Inhabiting forests, though not exclusively so, they are diurnal and gregarious, hunting in packs of six to twenty. They live largely on wild pigs and various kinds of deer, such as sambur and spotted deer, also Indian antelopes and even nilghai. In Tibet they feed on wild sheep. These animals are savage and extremely brave; they have even been known to kill a tiger. Generally they avoid the neighbourhood of man and rarely attack domestic animals, though occasionally they will pull down a domestic buffalo. They also eat vegetable foods, taking with apparent relish – and not merely medicinally as with dogs – herbs, grass and leaves.

These so-called dogs are practically untameable and have a rank foetid odour. They breed in the winter and from two to six or more pups may be born in a litter from January to March. The females make their nests in caves or hollow spaces among rocks. Sometimes several females associate together to form a breeding community. Dholes howl at night, but are silent when hunting; occasionally they snarl at each other, but do not fight.

Cyon albinus (*Northern Dhole*)

The northern dholes are found in north Asia and are rather doubtfully separated from *C. javanicus* by the larger size of the second upper molar. They live in mountains densely covered with forest and only exceptionally come out on to the open steppe. The packs appear to be very local and consist of 10 to 15 animals led by strong, fully adult males. Just occasionally they hunt by themselves, but whether alone or in a pack they prey on deer. These animals are both cunning and fierce and are greatly feared by hunters, who sometimes disappear into trees when dholes are in the vicinity. Dogs, also, dread dholes and turn back from traces of them as if a tiger were there.

The Bush Dog

Icticyon (*Speothos*) *venaticus* (*Bush Dog*)

The bush dog is a very curious canine type. It is believed to be of considerable antiquity and, as mentioned previously, remains have been found in caverns with Pleistocene deposits in Brazil. These animals are remarkable for the shortness of their limbs and ears and for their very short tail and muzzle. The body is relatively long and so is the neck.

They are omnivorous, but prefer meat to vegetable food. They are very bold and determined and dislike confinement. It is believed that they hunt in packs by scent and they can be very savage. Bush dogs are rarely seen. They have the characteristic that they take readily to water, though they never frequent the low-lands by the coast.

The skull is noteworthy for the shortness of the muzzle, which has a swollen appearance between the anterior margins of the orbits. The teeth are remarkable for the absence or minute size of the second upper molar in both jaws.

The Hyaena Dog

Lycaon pictus (Hyaena Dog)

This peculiar group of dogs is so-called because of their markings, which somewhat resemble those of hyaenas. They are separated from other dogs by the absence externally of both pollex (thumb) and hallux (big toe). They hunt regularly in packs, particularly at night, though they are often seen also by day. They run extremely rapidly and attack sheep, and cattle, usually trying to surprise them when they are asleep. Although they hunt in packs, they do not have a leader like the wolves and their attacks are somewhat unorganized. They never bark, but utter a shrill sound peculiarly their own. They hunt by scent and by sight. They do not make burrows and are not easily tamed.

Hyaena dogs attain the size of a tall greyhound, though their limbs are rather long compared with other Canidae. The head is broad and flat with a rather short muzzle; the ears are large and of a peculiar shape. The skull is short and thick and has a swollen appearance somewhat resembling *Cyon*.

The Large-eared Cape Dog

Otocyon megalotis (Large-eared Cape Dog)

The large-eared Cape dogs, known also as 'bat-eared foxes', are found only in eastern and southern Africa. They are the most aberrant of all the Canidae, both on account of the number of teeth and as regards their general proportions. The lateral aspect of the skull is very different from all other Canidae. They are the size of a large fox but stand higher on the legs and have shorter, though equally bushy tails. The ears are very large, like *C. zerda,* but relatively broader.

Appendix II

Variations of the Skull and Dentition among Wild and Domestic Canidae (plates 36–40)

By making careful measurements of the skull and other parts of the skeleton, many workers in the past have attempted to define species and breed characteristics by which the Canidae might be classified. Such efforts have all ended in failure and, if our arguments have been followed, the reasons should now be clear.

We have studied numerous skulls of wild and domestic Canidae, only to find gradations between the different groups and breeds; there are no hard and fast dividing lines. Clearly the characteristics of domestic dogs have radiated from certain atavistic traits present in the wild Canidae from which they are derived.

By means of selected pictures illustrating the most important features in the many skulls studied, an attempt has been made to show the main lines of development. Comment has been kept to a minimum, in the hope that the reader will study this material and discover for himself the grounds on which our main arguments are based. In essence these are that the genus *Canis*, whether wolf, jackal, fox, or domestic dog, is an entity within a single polymorphic genetical system. The fundamental traits developed in domestic dogs are all part of the general genetical inheritance, although some have become more pronounced in certain members than in others. Finally, although domestic dogs have probably been developed independently from different groups of wild Canidae at least four times, the main lines of development have been rather similar. The two main lines were those of the long-muzzled and short-muzzled dogs. Long-muzzled breeds have appeared certainly twice, from the northern wolves (sheep dogs) and from southern wolves (greyhounds).

Unfortunately, in only a few of our specimens is the sex known for certain and we have therefore omitted it in all cases. This is of little importance, since sex differences are not marked in Canidae. The two skulls of northern wolves from Whipsnade which show such marked dimorphism are both from males.

Appendix III

The Windmill Hill Dog

An entire skeleton of a dog and a number of other skeletal remains were recovered during excavations of Neolithic sites at Windmill Hill, near Avebury in Wiltshire. We were permitted to study these remains by the kindness of Major and Mrs A.A. Vatcher, joint curators of Avebury Museum. A photograph of the reassembled skeleton is reproduced on plate 39 facing p. 81.

The Windmill Hill people were primitive agriculturalists and pastoralists, and the deposits from which these remains were obtained are dated *c.* 3000 BC. The people grew primitive strains of barley and Emmer wheat. They kept small cattle (*Bos longifrons (brachyceros)*), sheep or goats, and pigs. Remains of red deer (*Cervus elaphus*), found in the settlement, show that they also hunted.

The skeleton is of a rangy type of dog, somewhat tucked in at the abdomen and with a deep chest, in this way somewhat resembling a greyhound. The tail, too, is thin, long and straight, without curls or bends, and in this respect it also resembles that of a greyhound.

The skull is narrow and the width between the zygomatic processes is also narrow; the zygomatic arch is well below the level of the top of the cranium. There is a marked 'stop', and the muzzle is relatively short. These features differentiate this dog from both greyhounds and collies and suggest an animal of the mastiff group. This impression is reinforced by the wide nares (nasal openings) of a scent-hunting hound. There is a well-marked sagittal crest.

The feet are wide and splayed, unlike those of greyhounds. The hind legs are well curved, not straight as in the dingo group of dogs.

The third incisor of the left ramus of the mandible is missing and the socket is filled with cancellous bone, indicating its loss during life. Incisors nos. 1 and 2 of the left maxilla are also missing, and the sockets are partially filled with cancellous bone. The two maxillae have become separated at some time during life, as is shown by the presence of 'exostosis' – new bone formation of healing.

The skeleton is of an adult dog, as is shown by the fused sutures of the cranial and limb bones. All the teeth are in very good condition, without undue wear from hard food and without signs of caries or periodontal disease. The tooth formula is standard for *Canis familiaris*. The wear on the teeth indicates quite a young dog, but it is evident that the animal received an excellent diet and appearances of tooth wear might be deceptive for this reason. The carnassial teeth are long and equal in length in the upper jaw to the two molars. The latter are small, but this condition might suggest a primitive form of domestic dog.

The mandibles are curved anteriorly, straightening a little posteriorly. A skull from a second dog rested on the mandibular processes, but needed only a light touch to make it rock like a wolf's skull. This again may be a primitive feature.

The second skull is of a smaller dog, though both were adult. Possibly one was male and the other female. Otherwise, remarks made about the first specimen apply to the second also.

The bones of the skeleton are in good condition and there is no sign of trauma or disease, apart from the condition of the mouth and the missing teeth.

One further mandible was seen; this belonged to a much larger dog and its silhouette is similar to those of the other two.

In addition to these three specimens, a skull cap of a very small dog was examined. Sutures are united and it was, therefore, adult. The head is rounded and there is no sagittal crest. A further skull cap of a very small puppy was also seen. These specimens were reminiscent of a pekingese or a pug.

The entire skeleton was recovered from a ditch, where it had evidently been thrown after death. This circumstance suggests a general purpose dog used for herding and hunting (a house dog would surely be buried). The injury to the jaw suggests a kick in the mouth from some animal, perhaps one of the cattle. An animal so obviously well fed, although thrown out after death, suggests that it was well cared for, and the mouth injury was for this reason unlikely to have been inflicted by a blow from its master.

From its long legs and build, the dog obviously had a good turn of speed, and the deep chest suggests stamina. The tentative conclusion reached is that this dog was of the prototype 'pointer' breed, which we have mentioned earlier.

The very small cap, lacking a sagittal crest, appears to belong to a different breed of dog, probably kept in the huts and fed on soft food. Possibly this is an example of the breed kept by the lake dwelling Neolithic peoples and assigned the variety name of *Canis familiaris palustris*.

Bibliography

1 ASDELL, S.A. (1966) Dog Breeding. *London: Churchill*
2 ASH, E.C. (1927) Dogs: Their History and Development. *London: Benn*
3 BACON, E. (1963) Vanished Civilizations – Forgotten Peoples of the Ancient World. *London: Thames & Hudson*
4 BIRD, C.G. & E.G. (1937) The Management of Sledge Dogs. *Polar Records* **3**: 180–4
5 BRANDER, M. (1964) The Hunting Instinct. *Edinburgh: Oliver & Boyd*
6 BRICE, M. (1934) The Tale of Your Dog: His Origin and his Needs. *London: Heinemann*
7 BRODERICK, M. & MORTON, A.A. (1902) A Concise Dictionary of Egyptian Archaeology. *London: Methuen & Co*
8 CAIUS, J. (1570) De Canibus Britannicis.
9 CAMPBELL, J. (1960) The Masks of God: Primitive Mythology. *London: Secker and Warburg*
10 CARRINGTON, R. (1963) A Million Years of Man. *London: Weidenfeld & Nicolson*
11 CLARK, G. & PIGGOTT, S. (1965) Prehistoric Societies. *London: Hutchinson*
12 COLE, S. (1965) The Neolithic Revolution. *London: Trustees of the British Museum*
13 COLLINGE, W.E. (1896) The Skull of the Dog: A Manual for Students. *London: Dulan & Co*
14 COMPTON, H. (1904) The 20th Century Dog. *London: Richards*
15 COON, C.S. (1963) The Origin of Races. *London: Jonathan Cape*
16 CORNWALL, I.W. (1964) The World of Ancient Man. *London: Phoenix House*
17 CORTES, H. (1520) The Despatches of Hernando Cortes Addressed to the Emperor Charles v. Tr. George Folsom (1843) pp. 112–13. *London: Wiley & Putnam*
18 H.H. DALAI LAMA OF TIBET (1962) My Land and My People. *London: Weidenfeld and Nicolson*
19 DE RACHEWILTZ, B. (1960) Egyptian Art. *London: Hutchinson*
20 DIAZ DEL CASTILLO, B. (1568) The True History of the Conquest of Mexico. Tr. M. Keatinge (1928). 2 vols. *London: Harrap*
21 DIGHTON, D. (1921) The Greyhound and Coursing. *London: Grant Richards*
22 DIXEY, A.C. (1931) The Lion Dog of Peking: being the Astonishing History of the Pekingese Dog. *London: Peter Davies*

23 FIENNES, R. (1964) Man, Nature and Disease. *London: Weidenfeld & Nicolson New York: Signet Science Series*

24 FERNANDEZ, J. (1965) Mexican Art. *London: Spring Books*

25 FITZPATRICK, P. (1920) Jock of the Bushveld. *London: Longmans, Green*

26 FRAZER, J. (1922) The Golden Bough. *London: Macmillan*

27 GREY, T. (1961) The Popular Chihuahua. *London: Popular Dogs Publishing Co*

28 HARTING, J.E. (1880) British Animals Extinct Within Historic Times, with some account of British Wild White Cattle. *London: Trubner & Co*

29 JENNISON, G. (1937) Animals for Show and Pleasure in Ancient Rome. *Manchester: University Press*

30 JONES, F.WOOD (1924) The Mammals of South Australia. *Government Printer, Adelaide*

31 KALESKI, R. (1946) Dogs of the World. *Sydney: William Brooks & Co*

32 KAMINSKI, M. & BALBIEZ, H. (1965) Serum Proteins in Canidae: Species, Race and Individual Differences. In Natousek, J.[Ed.]. *Blood Groups of Animals.* The Hague, Dr W.Junk, for the *Czechoslovakia Acad. Sci.*

33 KELLER, O. (1913) Antike Tierwelt. *Leipzig: W. Engelmann*

34 KING, H.H. (Undated) The Englishman's Dog in the Tropics. *London: Field Press*

35 LANE, C.H. (1900) All About Dogs. *London: John Lane*

36 LANE, C.H. (1902) Dog Shows and Doggy People. *London: Hutchinson*

37 LEE, R.B. (1894) Modern Dogs. 4 Vols. *London: Horace Cox*

38 LEIGHTON, R.G. (1952) The Complete Book of the Dog. *London: Cassell*

39 LOISEL, G. (1912) Histoire des Menageries. *Paris: Doin et Fils*

40 LORENZ, KONRAD (1954) Man Meets Dog. Tr. M.K.Wilson. *London: Methuen & Co*

41 LOXTON, H. (1962) Dogs, Dogs, Dogs, Dogs. *London: Paul Hamlyn*

42 LYTTON, N. (1911) Toy Dogs and their Ancestors. *London: Duckworth*

43 MENZEL, R. & R. (1948) Observations on the Pariah Dog. In Vesey-Fitzgerald, B. *Book of the Dog.* 968–99

44 MIVART, ST.G. (1890) A Monograph of the Canidae. *London: R.H.Porter*

45 MOORE, J.L. (1929) The Canine King: The Working Sheep Dog. *Cheltenham, Victoria: Standard Newspapers Pty*

46 MOWAT, F. (1963) Never Cry Wolf. *N.Y.: Dell Publishing Co*

47 PAYNE, R. (1964) The Triumph of the Greeks. *London: H. Hamilton*

48 PETRIE, W.M.F. (1892) Ten Years Digging in Egypt. *London: Religious Tract Society*

49 POCOCK, R.I. (1935) The Races of *Canis lupus. Proc. Zool. Soc. Lond.* **3**: 647–86

50 POCOCK, R.I. (1941) Fauna of British India. **1**: 84. *London: Taylor & Francis*

51 PRISMA, (1955) Encyclopédie Canine. *Paris: Prisma*

52 SCOTT, LORD GEORGE & MIDDLETON, J. (1936) The Labrador Dog: Its Home and History. *London: Witherley*

53 SCOTT, J.L. & FULLER, J.L. (1965) Genetics and the Social Behaviour of the Dog. *Chicago & London: University of Chicago Press*

54 SHAW, V. (1881) The Illustrated Book of the Dog. *London: Cassell, Petter Galpin & Co*

55 SMYTHE, R.H. (1958) The Mind of the Dog. *London: Country Life*

56 SOUTAR, A. (1929) A Chinaman in Sussex. *London: Hutchinson*
57 SOUTHERN, H.H. (1964) Handbook of British Mammals. *Oxford: Blackwell Scientific Publications*
58 STONEHENGE (1872) The Dogs of the British Islands. 2nd Ed. *London: Horace Cox*
59 SWEDRUP, I. (1958) Dogs of the World in Colour (Tr. R. Oldfield). *London: Blandford Press*
60 TALBOT, J.S. (1906) Foxes at Home. *London: Horace Cox*
61 TUDOR-WILLIAMS, W. (1954) Basenji: the Barkless Dog. *London: Watmoughs*
62 VESEY-FITZGERALD, B. (1948) The Book of the Dog. *London: Nicholson and Watson*
63 VEVERS, G.M. (1948) On the Phylogeny, Domestication and Bionomics of the Dog (*Canis familiaris*). In Vesey-Fitzgerald, B.[62] 1–20.
64 VLASTO, J.A. (1923) The Popular Pekingese. *London: Caxton Press*
65 WILCOX, A.R. (1963) The Rock Art of South Africa. *Johannesburg: Nelson*
66 WINGE, H. (1941–42) The Interrelationships of the Mammalian Genera. *Copenhagen*
67 WINGE, O. (1950) Inheritance in Dogs: With Special Reference to Hunting Breeds. *Ithaca: Cornell University Press*
68 YOUATT, W. (1845 & 1874) The Dog. *London: Charles Knight*
69 YOUNG, S.P. & GOLDMAN, E.A. (1944) The Wolves of North America. *Washington: American Wildlife Institute*
70 ZEUNER, F.E. (1963) A History of Domesticated Animals. *London: Hutchinson*

Index

INDEX